*"To my wife Indiana, who enabled me to complete this
work by nurturing me back to health"*

Cover illustration: Courtesy of **ESA, NASA/JPL-
Caltech**, and Felix Mirabel (French Atomic Energy
Commission and Institute for Astronomy and Space
Physics/Conicet of Argentina)

AETHER
The Physicalists' God

Contents

Life has been a quest going from J. L. Borges' "The Garden of Bifurcating Paths", to Everett's Many-worlds Interpretation of Quantum Mechanics. From the collective behavior of sardine schools and J. Cortazar's eels in "Prose from the Observatory", to Bose-Einstein condensates. From the relativity demonstrated by the twins paradox thought experiment, to the non-locality uncovered by the EPR (Einstein, Podolsky, Rosen) experiment. After thoroughly studying the physics of Nature, I am convinced that the aether is the physical, but immaterial substance, from which the universe emerged.

Introduction

Aether is the empty space in which the universe sits.

Empty space is real but does not exist as matter. Einstein was right, the universe is background free. The gravitational aether does not exist, yet it is the physical but immaterial substance from which the universe emerged.

> *"It need hardly be pointed out that with things that do not change there is no illusion with respect to time, given the assumption of their unchangeability."* (Aristotle, 340BC)

How big is a point? That is just like asking how big is empty space. Size does not apply, points are dimensionless. Same with empty space, it is dimensionless, yet contains the universe. This is why it is said that we could fit the whole universe in a point. In this realm, we need to think in terms of state, not in terms of process. Process occurs as spacetime. Trying to mathematically describe empty space using classical physics laws is not enough. Notions like size, age, velocity, or infinity, which imply motion, quantity, extension, or duration, should not be used to describe empty space. That is what Einstein meant when he said there was no time before the Big Bang. At the aether scale there are no distances to cover, it is all pervading... the aether is one. Instead of asking what happened before the Big Bang, it makes more sense to ask which was its state.

The gravitational aether is but does not exist as matter. It is before spacetime, or the Big Bang, or Inflation, or even Wheeler's Quantum Foam. It is not matter, therefore notions like motion, size, or duration are not applicable. Time does not apply. It is outside the rules of spacetime.

This notion of a primordial substance is a very old one, also known as Akasha, or Brahman, and many times described as pure energy, or spiritual fire. It has been anthropomorphized by man since the times of Plato and Aristotle, the Chaldeans and the Akkadians. Called by names like Zeus, Jupiter, Brahma, and other. Always seen as immaterial, until 1964, when the cosmic microwave background radiation (CMBR, CBR or CMB) was discovered. Since then, there have emerged completely contradictory notions which now compete for acceptance. The reductionists are becoming restless in countless desperate attempts to quantify the unmeasurable. Now there are new claims of an absolute frame of reference showing up everywhere. They claim they finally have a fixed inertial frame, as if we ever needed one. But according to Relativity, objects in spacetime are relative to one another, not to empty space.

Contemporary physics is increasingly turning into a cross between General Relativity and Quantum Mechanics. After we realized there was Wheeler's Quantum Foam, we have slowly integrated particulate space into GR, while starting to take a serious look at emerging Space Flow theories combining QM (Quantum Mechanics) and Relativity.

Here, I use the aether concept, Quantum Mechanics, and Relativity in an attempt to solve what many philosophers call the Hard Problem, and to answer: *What is that which is?*

God is a Thing, Not a Person

In most aether views, whether material or immaterial, the aether is seen as an all knowing creative force, but not in this view. I compare the aether to God in the sense that it is one, omnipresent, and eternal, but at the same time I argue that it is not all knowing, that it is a thing which can neither think, nor see without a brain. That it sees, thinks, and exists through matter.

> *"A substance cannot be produced from anything else: it will therefore be its own cause, that is, its essence necessarily involves existence, or existence appertains to the nature of it."* (Spinoza, 1673)

Basil Hiley is correct, being remains constant during the process of becoming. Matter is only temporary, it has a beginning and an ending, it is subject to time (change). Things are because of the aether, it is what gives them their *temporary being* status. The Real, as Hegel called it, *that which is*, is the aether. Reality, on the other hand, simply refers to the process of becoming. The aether becomes through matter.

Energy is finite, as you probably know, this is the reason why nothing with mass can reach the speed of light. As a proton reaches the speed limit, its waves are flattened, it loses its wavelength and goes back to being aether. Slow down the system and it reappears... as required by local spacetime conditions. For a single proton to reach the speed of light, it would require more than all the energy available to the universe. Because energy is finite and the speed of light needs to be kept constant for fields to work in the allowable speed range (0 to 300,000 km/s), there is time dilation and space contraction for material systems moving at relativistic speeds.

> *"This shows us two things: you cannot have parts of the infinite and the infinite is indivisible. But indeed, even if the One is more like a Principle, and the one is undivided, then the whole universe will be undivided either in quantity or in form."*
> (Aristotle, 340BC)

Matter is 99.9999...% empty space. Reduce yourself to the size of the smallest particle and you will still see nothing more than empty space. Matter is made of fields and fields are little more than apparitions. Fields are shapes in empty space, lines of force. Matter seems like an illusion, but that is reality, and matter in spacetime is the one drawing the shapes, not some creator or designer. The universe designs itself. Particle creation occurs according to local spacetime's energetic or thermodynamic requirements. Reality is process and process happens as spacetime.

Matter is condensed space and energy is space in motion. Matter is the same as energy, hence $E=mc^2$. Whether there is an aether or not is finally answered: the aether is but does not exist until it turns into matter.

This is not a new theory but a new insight into already existing theories. A freshly synthesized interpretation, consistent with already known and well accepted scientific facts. A modern perspective in which the aether concept is reintroduced in an attempt to reconcile a centuries old notion of wholeness in space and time with actually established scientific paradigms. In addition to arguing for a common substrate to all matter in a purely dialectical way, without math or complicated formulas, I relate self-awareness and perception to non-living, self-organized systems. Thereby suggesting that Consciousness is not an independent supernatural entity separate from matter, but an intrinsic property of all matter.

We need to take a better look at the cold hard facts. We have made huge technological advances, but spiritually and therefore politically, we have been stuck in time for the last two thousand years. For Humanity to solve its spiritual and political problems, it needs to conclusively figure out the relationships that exist between being, matter, and space. Before we can move forward, we need to make up our minds between fantasy and reality, we have to choose between superstition and reason.

This book represents, in a short and informal style, what I have realized after a lifelong quest for proof of wholeness in space and time as a fundamental property of the universe. It is aimed at a general audience, going from the specialist to the layman, with the hope of further popularizing these deeply philosophical issues.

[If you read all the chapters in a row, it may feel repetitive, but this better enables the reader to read any chapter independently from the others.]

Aether... where Physics meets Philosophy

Leading cosmologists picture the universe as a bubble floating in empty space and Einstein's spacetime as the space inside that bubble. Now, is that empty space composed of parts? No.

Do the concepts of motion, and therefore time, apply to it? No.

Does it have a beginning and an ending? No, it does not move, and therefore not subject to change. It is eternal.

Is it everywhere? Yes.

Is it the seat to all fields? Yes.

Can there be matter without fields? No.

Is it matter? No.

Is it real?

Traditionally, Western science's tendency has been to fragment and isolate everything we take as the object of our investigations, ignoring the background or the underlying substrate from which the universe emerged.

From the Copenhagen Interpretation of Quantum Physics and the Heisenberg Uncertainty Principle, we gather that light is particle and wave at the same time. That the totality is more than the sum of its parts, and that, when you get down to the size of atoms, there are no solid-like particles spinning in empty space, but a net of interconnected wave-particle systems: a hologram ruled by the laws of Quantum Mechanics, Relativity, and Thermodynamics. From the EPR (Einstein, Podolsky, Rosen) experiment we find that, regardless of the distance between the two, when we measure the spin of one of the photons on a pair of entangled photons the other photon registers the spin direction instantaneously. Which gives us non-locality at the quantum level. And, from John A. Wheeler's Delayed-Choice and John Cramer's Transactional Interpretation of Quantum Mechanics, we get undividedness of process, wholeness, self reflection, and self-organization.

From these facts we can argue that matter originates at a deeper level, and that state is instantaneously registered throughout space thanks to wholeness in space and time. This wholeness, I believe, is what makes these phenomena possible.

It has been argued before that there is an interaction at a deeper level between matter and the environment in which it develops. This notion that energy and matter come from a common substrate is a very old idea. And that is the aim of this book, to examine the philosophical implications that these new scientific facts bring to light, and to reassess this new state of affairs.

I argue, as many others do, that the aether is the physical but non-material substance from which the universe came to exist. To exist, things must be in spacetime, but the aether is not in spacetime, it is before spacetime. It is, but does not exist as matter.

The aether is all permeating, it is inside and between particles, it is everywhere by its own definition. Everything is made from it, even the space that surrounds you. You cannot conceive a fragmentable aether, or it would not be the aether as it was defined thousands of years ago. It is indivisible. Fields can create the appearance of separated volumes, but you cannot divide the aether into separate entities. In that sense, it is apparently, infinitely divisible.

The aether, as described over four thousand years ago, is materially non-dimensional. It is not matter, therefore not directly observable. You can measure the properties of fields, but you cannot take a direct measurement of the aether.

Motion is not one of the aether's properties, neither is time, nor change. This makes it immutable, or eternal. Since it lacks the property of motion and cannot be described as containing parts that follow a time-line, we can conclude that it is not matter. At the sub-quantum level, the level at which energy is before it turns into multiple entities, motion loses meaning. Any material substance will occupy space, but this physical non-material substance does not. It becomes matter as fields vibrate, or pulsate at very high speeds. Creating material properties like volume, extension, motion, time, mass, gravitation and solidity, eventually causing the formation of objects in spacetime. Once we have the limits, the boundaries, we can talk about notions like size, extension, motion, time, and process.

In this view, the aether has no capacity to hold any *active information*, just *passive information*, the constants, which are used by active information as energy is turned into quantities, or quantized... in spacetime.

The aether gives the universe properties like wholeness, interconnectedness, continuity, and non-locality. There are no parts when you refer to the aether, but you can look at electric and magnetic fields as different *things*, or *parts* of a greater whole. All made from the same continuous and non-fragmentable aether. Everything is connected to the aether because everything is made from it. This is where wave-particle complementarity comes from.

No Mind-Body Gap

According to contemporary Quantum Mechanics, wave-particle complementarity is due to an indivisible process which originates in a common background, but it appears as if the only necessary information being transferred through EMR (Electromagnetic Radiation), from the aether to the particles, is that concerning the momentum and location of the particles in relation to that inertial frame and the rest of the universe. The rest of the information needed for the evolution of the system in spacetime is contained by the system itself, in spacetime. Thereby eliminating the need for an all knowing god, or creator.

$[X, P] = 0$ --> commutativity (leads to a dualism)
$[X, P] = ih$ --> non-commutativity (leads to a monism)

Hence the non-separability of process claimed by so many.

> *"The non-commutativity of the underlying process produces an ontological complementarity. This must be contrasted to Bohr's epistemological complementarity."* --- Basil Hiley

Contemporary Quantum Field Theory supports the idea that the ontology is in the process which matter undergoes as it fluctuates in and out of nothingness. Classical Physics being a description of what the world appears to be, and Quantum Field Theory a description of what the world is.

According to Louis de Broglie, et al., every object exists as a body coupled to a matter wave, or pilot wave, and its displacement through space can be described by a wave-function. Information about the object's relation to its surroundings and the rest of the universe is picked and brought in by each object's particular pilot wave.

Bodies in motion need to continuously reset their energy requirements. As we now know, particles are not these space independent billiard ball-like objects floating in space, they are wave-particle systems in constant motion. Cloud-like standing waves which require a continuous energy flow from the substrate to the particle. This is why position and momentum cannot be known at the same time. This is where the Heisenberg Uncertainty Principle comes from.

The Real

Matter is continuously changing, becoming. What was a second ago is no longer, and the only things real or meaningful to us are the information and processes through which things become and now are. But the immutable, the eternal, the real, is the empty space in which the universe sits. Matter and fields are little more than apparitions, active information, as David Bohm called it. Basil Hiley, one of David Bohm's collaborators, is correct when he says that *being is a relative invariant in the overall process of becoming*. The fundamental laws, that which remains unchanged, is what is real.

Can you be without materially existing? Logic tells us that creation ex nihilo is physically impossible. And from electromagnetic phenomena and gravitation we get that, physically speaking, to be, you don't need to be material, all you need is to be able to act as a force. Existing is not the same as being, you can be without existing, but you cannot exist without being.

Is empty space real? Can we prove it? Can we measure it? Can we mathematically describe the rotation or acceleration of an object in empty space without assuming empty space to be real? I mean, if you were the only particle in space, how could you tell when spinning or accelerating? Is the only time we can have space, rotation, and acceleration when we have more than one object to consider? Empty space may be empirically untenable, but it is already considered as real by present theory. Which is why we have spacetime metrics.

The aether is not in spacetime, spacetime is in the aether. Empty space and spacetime are not the same thing. Einstein's spacetime is material, empty space is not. There can be no space without time nor motion, this is why Einstein called it spacetime. As Einstein once said: if we had no time (process), everything would have to happen at once. That is why Einstein described reality as a spacetime continuum where he saw process as the weaver of the *fabric of space*, a fabric made from space and time. Reality is process... spacetime is process.

Time, space, and matter start with the quantum, and quanta can exist only when in motion. Field motion, or energy, turns into matter. If we could stop the motion, matter would go back to being just empty space. Outside of time, quantum events are not possible. There is time and space because there is motion, and there is motion because there is energy. The aether itself does not move, matter does, the quantum does. As Einstein used to say: *energy is space in motion*. In this sense, aether is synonymous to energy, it is pure energy.

In this view, the aether is the substrate to all matter, including Wheeler's Quantum Foam. It is before geometry. Everything depends on this substrate, this is where the laws of gravity and electromagnetism are administered from.

Electromagnetic fields should not appear as ultimate, irreducible realities. Existence starts with the field, and before that there is what we call empty space, or aether, which is neither big, nor small: extension is not one of its properties. Spacetime and geometrization happen after the aether. The aether, unlike spacetime, is primary. Matter, space, and time are not. Empty space which is not really empty but full of pure energy. Energy which exists before EMR, and therefore is neither hot, nor bright.

Aether and Relativity

Empty space was, then came the universe. There can be empty space without a universe, but not a universe without empty space... just as there can be no matter without continuity, causality, or process.

Relativity is an aether theory. Everything is related through empty space. That everything exists in a single field, a matrix, is the basis, the foundation of the General Theory of Relativity. It is a philosophical necessity, if not, how could anyone explain why we have inertia, gravity, or action-at-a-distance?

Thanks to Mach and others, Einstein saw that we needed a metric, that we could not accurately describe reality without taking into account what each body in the metric was doing. That in order to explain events according to fact, each point had to be connected to all the other points. That without a metric and relativation, concepts like gravitation, rotation, acceleration, and inertia could not be clearly explained. All he had to do to explain Relativity without an aether was explain motion as a function of position determined through the metric. Using tensors which tell matter where to go at each point in the metric. What this meant is that the trajectory any body in space takes would depend on its position in relation to its surrounding objects. Why? Because space is flowing into each and all the surrounding bodies, giving each point in the metric a direction and a force. Which is where tensor math becomes very useful.

But what else is this metric if it isn't a mathematical representation of the all pervading and unifying aether? Einstein's intuition and common sense told him that, in order to explain inertia and gravitation, there could not be this bunch of separated and unconnected bodies (Newton's Billiard Ball model). This notion of a gravitational aether, a continuous field from which spacetime emerges as objects interact with each other, is the philosophical basis for the General Theory of Relativity.

He also realized that an absolute frame of reference would invalidate Relativity. That by being relative to a fixed inertial frame, things would not need to be relative to each other and there would be no relativistic effects, which we already know is contrary to the facts. There is no need for an absolute frame of reference when you have an infinitely divisible substrate from which everything is made.

This notion of an aether has been integrated into physics for a very long time. We are talking about the perfectly flat vacuum state of Quantum Field Theory. A perfectly flat vacuum state which refers to the quantum mechanical state of the vacuum. But is this vacuum considered a thing? Is it real even though it is not matter? I suppose it is, since how could it be in any given state if it wasn't real? It is real and it is called aether. Some call it the long winded vacuum state of quantum field theory, I prefer to call it what it has always been called: aether. [Since it harkens back to the idea of a fixed frame of reference, which can be misleading, the term aether isn't used much these days.]

Einstein's gravitational aether is the seat to an all relating process which he called spacetime. The aether was re-introduced early in the 20th century by scientists like Einstein, Mach, and Minkowski as they were trying to describe a substance, or... a thing. Einstein said that matter and fields emerged from the same basic substance, that there could be no universe without an aether because it is the seat to the electromagnetic and gravitational fields. That there are gravitational, electromagnetic, and nuclear forces because there is an aether. According to him, without fields there can be neither matter, nor spacetime, therefore the aether is.

Einstein's Universe is Background Free

But this is not the same aether Newton, Poincare, and Lorentz talked about. Particles come to existence as required by spacetime's energetic and thermodynamic conditions. In this new aether, objects are relative to each other, not to absolute space, therefore there is no violation of the Principle of Relativity. As they explained Relativity, Einstein, Mach, and Minkowski said that things are not relative to absolute space, but to an absolute world. Acceleration is measured in relation to other objects in spacetime, not in relation to absolute space. According to Mach, this is why there is inertia.

Einstein's gravitational aether does not represent an absolute inertial frame. It is not material, therefore cannot represent a background. It is not quantized, like material space. Einstein was correct in his claim of a background free universe in the sense that there are no landmarks to be used as reference to motion, or elapsed time. How could a non-material aether represent a preferred inertial frame if it lacks any landmarks or coordinates? It cannot.

Einstein, Minkowski and Mach described a different aether. This twentieth century aether differs from earlier aethers in that, in it, objects are relative to other objects, not to empty space, therefore avoiding a Principle of Relativity violation.

As Einstein said, spacetime is an extension of matter. That is because spacetime and empty space are not the same thing. Spacetime is neither primary, nor fundamental, it does not exist by itself, it is a product, just as matter and time are. There is flat empty space, then there is curved spacetime, or what is known as the *observable universe*.

Therefore the universe is background free and there is no fixed, nor absolute frame of reference. There is absolute reality. From Einstein's General Theory of Relativity we get that objects are not relative to empty space, they are relative to other objects with mass. Respect to Relativity, what is absolute is not empty space, what is absolute is the objective universe, the world. This is what makes GTR (General Theory of Relativity) true, everything is related through and by the aether. Or, how could it be that when a body is accelerated to near the speed of light, time and length must change in relation to a stationary observer? Wasn't space supposed to be absolute, primary, independent, and not derivable from anything else? According to General Relativity, the universe is one single entity, one process. Space... objects... Mankind... all come from one thing, which by definition, we call aether.

Einstein presented a different notion of the universe with his 1920 essay *Ether and the Theory of Relativity*. He stripped 19th century aethers off any kinematic or mechanical properties. This new aether lacked the property of motion and was not composed of parts which followed a time-line. What he termed the *Gravitational Ether* came from a completely different idea. Motion and particulation, he said, cannot be considered properties of the aether because it is one and has no components. This oneness can be used to explain action-at-a-distance, gravity, and inertia.

Einstein's aether is more akin to Newton's absolute space than most people think, this is why he sees the universe as background free but imbued with Mach's reciprocity between matter and space. It is Newton's absolute space mixed with Mach's aether, or with relativation. Empty space tells matter what to do and matter tells empty space how to curve. Which is where space curvature comes from.

Einstein said that, when trying to define the aether, we need to put aside notions of motion, extension, size, beginnings, and endings. In essence, he said that this substance lacks the properties of matter, yet all matter emerged and is ruled from it.

Spooky Action at a Distance?

Not Really. Bodies in space never acted on each other from a distance, as Newton argued while explaining gravity: there is no action at a distance because there are no distances to be covered. The aether is one and everywhere, it has no moving parts, motion is not necessary. This is why state can be instantaneously registered throughout material systems. Which means that there are no faster than light (FTL) information transfers, just changes in state (where stress-energy tensors and lines of force can be affected), at the aether level.

If the aether is an all-pervading substance, why would it need a property like motion? Motion and time are for objects, for parts which follow a time-line in spacetime. It is an error to think in terms of spatial extension when trying to understand what is going on at the aether scale. The aether is everywhere, it is the set of all sets. It is the circle Zeno, Bruno, St. Augustine, Pascal, and Borges among others, once talked about; a circle whose center is everywhere and circumference nowhere. The aether is not dependent on geometry, but helps determine the geometry of spacetime. It is a plenum, a matrix... the origin.

This notion of wholeness is probably what triggered Einstein's interest in Bohm's *Undivided Wholeness*. He understood that for there to be a continuum and Relativity to hold, the universe must be conceived as a whole. Why else would an object's dimensions depend on its surroundings if it wasn't for this wholeness? [He eventually became a Pantheist.]

Relativity can only refer to relative time or length because it is the description of a whole where objects are physically and energetically dependent on each other. In other words, if an object were to be conceived as accelerating at relativistic speeds in a perfect void, independently from any object or frame of reference, there would be no time dilation, nor length contraction. But because objects are embedded in a continuous field, a metric which represents the whole, and because the whole's energy is finite, objects exhibit relativistic effects in relation to other objects. It is a property of the whole which arises from a physical need to abide by the laws of Thermodynamics.

The Fundamental Forces

Some claim that empty space has no physical properties, but if you eliminate the notions of permittivity and permeability from Maxwell or Einstein's equations, ratios on which the existence and behavior of all fields entirely depends, the theories will completely fall apart. Some believe in the reality of nothingness, that empty space as such is real, and accept that notion as an integral part of their physics, but can't even ascribe any physical properties to it. At least Einstein's aether can be said to be real because of its physical nature. It is physical because it can act on matter, and immaterial because it lacks properties like extension and motion. It does not move and has no parts or components in the material sense. To be real there is no need to be in spacetime, just to be able to act in spacetime.

When you have a magnet acting upon an object we say that a magnetic field is what moves the object. But the path, the direction of propagation, and the strength of the magnetic force lines are determined at the aether level.

Free space (empty space) is where the laws that

21

determine the behavior of the four fundamental forces of Nature are set. Laws by which active information (matter) exists. The aether holds basic, changeless, motionless, fixed values; hidden dimensions from which everything else, including meaningful information, is being formed as spacetime.

As many have said, the aether helps determine things like the ratio between the electric displacement and the intensity of the electric field producing it, in free space (permittivity). Or the ratio between the magnetic flux density and the external field strength, in free space (permeability).

The aether itself is not observable, you cannot say - *here, lets take a look at this piece of aether!* Because it is not matter. It is real, but not in spacetime, hence not directly observable. This is why the MMX (the Michelson-Morley experiment) failed so miserably. But you can measure its effects: things like inertia, gravity, magnetism, electrical charges... etc.

From the MMX results we should conclude that the aether is immaterial and directly unobservable. Now, if there was an empty space before there was matter, isn't the classical vacuum also immaterial and directly unobservable? Can we take a direct measurement of something which is not matter? The only thing proven by the MMX was that we did not understand the aether's nature. Want to measure aether caused drag? Just measure a moving object's momentum, or measure the force needed to accelerate any object, that is aether caused drag!

Is Empty Space Curved?

Some say, but if there is no material aether, how come spacetime is curved? It is curved because of sidereal lines of force. What is curved are the lines of force flowing space follows as it is sucked into bodies with mass. Spacetime is warped as space sucking bodies with mass are added to it. All particles being affected by those lines of force as they move through spacetime (i.e., gravitational lensing). Mass (process) is what causes spacetime curvature, this is why it is not flat. This is why we need a non-Euclidean geometry to describe it. This is where tensor math and space curvature come from. Empty space has physical properties, which is why tensors are needed in order to determine the forces involved in the process and the trajectory particles will take as they move through it.

Laniakea Our home supercluster.
See how all matter, including space, flows towards the center of gravity (The Great Attractor).

Aether vs. Spacetime

Before we continue, we must further distinguish empty space from material space, or spacetime. EM fields and matter are observable, or measurable, empty space is not. EM fields have a geometric structure, empty space does not. When you describe an EM field, you may talk about intensity, density, size, magnitude, or duration, but not all of these concepts may be properly used to describe Einstein's gravitational aether.

I see empty space as the seat to EM fields, synonymous to Einstein's aether, and I see it as a primary, or fundamental component of physical reality. Material space, or what many call the Cosmic Background Radiation (CBR), remnant radiation left after the Big Bang, is seen as a product. Since, in this view, aether and empty space are synonymous, from now on I will refer to them as one and the same thing.

Einstein's aether is not the same as his spacetime. Spacetime is an aether product, synonymous to Timothy Boyer's material space, which is nothing more than a combination of CBR, Wheeler's Quantum Foam and invisible quantum matter. Spacetime is material, and Einstein's aether is physical but immaterial. First, there needs to be an aether before we can have anything like EM fields, matter, spacetime, or even Wheeler's Quantum Foam.

We have to be careful with meanings. What Einstein was referring to as *empty space* is more akin to a perfect vacuum than the space we usually talk about. Remember that, at the time Einstein wrote his 1920 essay, Big Bang and Inflation theories were still in their infancy. He thought that, in order to obtain an empty space, it was possible to extract all matter from a given volume. It wasn't until Timothy Boyer and what many call, the Casimir effect, that we began to really understand the meaning of empty space. In a true empty space, there can be no Casimir force.

Big Bang and Inflation theories may still need more work, these may still be incomplete. I am starting to think that matter pops up where matter is less dense, as if the universe were fighting against fragmentation, giving rise to objects like Quasars where it is less dense, and Black Holes where there is too much.

> *"The particle can only appear as a limited region in space in which the field strength or the energy density are particularly high..."* --
> - Albert Einstein

Particles can emerge anywhere and as needed, e.g., particle pair creation, but from where and what do they feed from, creation ex nihilo? I don't think so, that seems like a physical impossibility. Anyway, why would we have wave-particle complementarity if it were not because matter depends on the substrate? Isn't this the reason why we need a Higgs mechanism?

The aether has a non-zero vacuum expectation value, that is the reason why particles emerge as expected... or as dictated by spacetime conditions. This is probably why Alan Guth calls it a false vacuum.

The way I see it, what keeps the whirlpool moving in a spiral galaxy is space tension (the stress-energy tensor, the source for gravity, space curvature, and hence gravitation). Nature fighting against fragmentation, or against entropy.

We must not confuse the concept of space outlined by Inflation theories with Einstein's gravitational aether. Extension is a material property not applicable to Einstein's aether, neither is density. On the other hand, Einstein's spacetime is material, and properties like extension and density do apply. But this is something we learned after Einstein's Relativity. Inflation theory came after Relativity. Remember Einstein's cosmological constant, which he later called his greatest blunder? Einstein's universe was initially static, then, after Hubble, he learned about what we now call Inflation.

EMR, CMBR, and ZPE are all observable, material phenomena, with mechanical properties, like density and pressure. The universe inflates as background radiation, Wheeler's Quantum Foam and dark matter fill spacetime. Spacetime is packed full of particles.

Aether is Finite

The aether is a spatially boundless but physically finite substance. Energy is finite. Mass is finite. Finite, because if it were not, we would not be having phenomena like time dilation and space contraction, there would be no need for energy conservation. Zeno was right, the aether is infinitely divisible... but, as Relativity says, physically finite.

Imagine a totally empty space with no boundaries, what

do you have? You have Basil Hiley's pre-space. Spatial extension becomes an unnecessary concept. Since it lacks any landmarks to use as reference points, you are unable to measure extension or motion, thus cannot tell size nor distance. This is also why allowing for an aether does not constitute a violation of the Principle of Relativity. According to Relativity, things are not relative to empty space, things are relative to the universe. If objects were relative to empty space, all objects would have to move in relation to absolute space and time, and there would be no need to include covariance as we explain acceleration. But the reality is that we do need covariance to accurately describe objects under acceleration, especially when moving at relativistic speeds.

Since the aether's energy is finite, time and space will contract and dilate accordingly, while mass will increase or decrease. Reality automatically adjusting itself to present spacetime conditions and thermal requirements as matter follows Nature's fundamental laws. Because it is physically finite, matter and energy are also finite at any given moment, but infinite as a function of time and transformation. Even though proportions and ratios are kept constant, spacetime dimensions must be constantly adjusted to fit each inertial frame.

The reason that objects in spacetime must be related is because it all comes from a single entity, reality is one single process. All inertial frames within the observable universe are related by the aether, through the aether. The aether is one, and because of that, the universe is also one. It is a single entity, yet contains everything that exists in spacetime.

Einstein's aether is not bound by time but by topological properties, a set of ratios determined at the aether scale. Frame independent fundamental constants, a very small number of fixed laws by which all matter must abide. Physical (real) but non-material quantities (topological). Time independent continuity and connectedness. We can also call it topological space, a false vacuum, inertial space, or momentum space.

Lorentz invariant values originate at the aether level, they are real but non-material ratios. Hidden dimensions which help determine the geometrical properties of objects in spacetime. This is the level at which frame independent constants, like the propagation speed of fields are set. Take the fine structure constant for example, change its value and you may get a totally different universe. Thanks to these frame independent constants, the universe is isometric.

What did Murray Gell-Mann use to create his multidimensional geometric structure? A bunch of extra, hidden dimensions, or fundamental topological values that could represent reality. He learned that by manipulating this structure he could reproduce real world interactions. As he placed his extra dimensions into this new geometry, he found that he needed a few new particles to complete it, to fill the gaps, so to speak. He predicted what some of these fundamental particles would be before they were actually found, earning himself the Physics Nobel prize in the process.

Spacetime is four dimensional, and all these other hidden

dimensions rule matter but are not matter as such. This is why, in addition to analytical and non-commutative geometry, we need tools like topological quantum field theory (TQFT) to effectively describe what happens at the aether level.

If we could manipulate and rearrange its parts, and as long as we could maintain and continue to use the same fundamental constants, we could take a pound of earth and turn it into a pound of gold.

Because the aether is finite, and the propagation speed of fields must be kept constant, we need the equivalency and relativity principles. We need to describe reality in terms of covariance and deformation in order to be accurate. Matter changes, but not the fundamental values it follows as it takes shape.

Covariance and Deformation

Consider $c = 1/\sqrt{u_0 e_0}$, where u_0 and e_0 are permeability and permittivity in free space.

This relationship holds true because the speed of light (and of all electromagnetic phenomena) is determined at the aether level. It remains constant in all inertial frames because it is not dependent on a coordinate system, as matter with mass is. Since ratios like permeability and permittivity are determined at the aether level and the aether is immaterial and not bound by spacetime laws, 'c' (speed of light) can be frame independent.

The speed of light is frame independent, and only dependent on the physical properties of free space. This is why the speed of light is a constant unaffected by neither the speed of the observer, nor the speed of the observed.

The speed of light sets the scales. For fields to continue to work regardless of spacetime conditions, there must be time and spatial distortions between the observer and the observed when moving at relativistic speeds. This is where the principles of relativity and equivalency come from. Because a field's speed must not change regardless of relative motion, and because energy is finite, for reality to work, all parameters must be adjusted around the speed of light. This is how and why we get time dilation and length contraction.

Since the speed of light, hence the propagation speed of fields, must remain constant for all the other fundamental constants to continue to be proportionally the same, mass (process) has to increase in order to keep up, but to a limit. Once we go over the speed limit and fields can no longer keep up, matter disintegrates. When we reach the speed of light, wavelength and frequency drop to zero, waves become flat, devoid of any information, and we are back to being immaterial empty space.

Spacetime is dynamic. Time and length contractions are real. They need to be in order for the Equivalency Principle and the laws of Thermodynamics to hold. Spacetime, or material space, is a product, not a fundamental or primary component of reality, which is precisely what is claimed by Relativity. In spacetime, space-like separation is relative. If spacetime were static, as Newton's absolute space, then spatial extension would be neither variable nor dynamic, but it is, it shrinks and expands, just as clocks run slower or faster, depending on energy usage vs. energy available.

Since the speed of light is constant and closely related to the Compton wavelength and the Schwarzschild radius, the universe is the same everywhere, independently from existing spacetime conditions. This means that matter will always have the same properties and behave the same way everywhere, regardless of the existing spacetime conditions. It means that a carbon atom will look and behave as a carbon atom basically anywhere in the universe.

Space is Material

The Smallest Wave

Most scientists agree with Max Planck in that quanta are the smallest measurable energy amount or quantum of action, and that Planck's constant 'h' relates the energy in one quantum of electromagnetic radiation to the frequency of that radiation. That all of matter's spectral properties and patterns can be explained in terms of exact multiples of a basic minimum value, and that thanks to the determination of 'h' we can now conceive order at the quantum mechanical level.

Many also agree with Louis de Broglie in that every object in motion would move in a wave and is accompanied by a wave. Also, that he showed that electrons traverse an integral number of wavelengths for each complete orbit of an atom, and that the quantization and structure of quanta will always depend on the properties of their source and the objects with which they interact.

Quanta are measurable amounts of cycling spacetime. In reality, there are no more spinning particles than there are wavefronts. A quantum of action (quanta) appears as a very small region of fluctuating, process independent space.

There are no point particles. A point particle must be internally static, with no internal time, and that is not possible. Points do not occupy space, they do not exist. Which is why the physicist refers to these as 'point-like' particles. The point being just a mathematical convenience.

Each particle, even the smallest, is in constant internal motion, vibrating. Each particle possessing its own internal time and wavelength. For this reason, it is not possible to get rid of uncertainty at a spacetime level. This is one of the reasons why sometimes we need to round up measurements to no less than 4 decimal places in order to more accurately describe reality.

Planck's constant is closely related to a particle's wavelength. It is from this relationship that wave-particle complementarity and the non-commutativity between momentum and position emerge as properties of space and matter, and as proof of the inseparability of quantum processes. Momentum and position do not commute because subatomic particles cannot be seen as standing and moving at the same time. Energy is constantly flowing in and out of the system as matter continuously feeds from empty space.

The fundamental particle should then be described as a small spherical region of pulsating spacetime whose diameter equals one Planck length (10^{-33} cm). And space can be described as packed-full of particles, or as a sea of randomly fluctuating particles. This is why space is considered by today's physics to be grainy quantum matter, each grain having its own internal motion, or cycle. Some having a diameter as small as one Planck length, billions of times smaller than any neutron or proton.

Each object, regardless of size, is accompanied by its own particular matter wave. Therefore the concept of a superwave function, where various particles are described by a single wave-function, applies.

Gravitational Aether

The aether I believe in is akin to Basil Hiley's pre-space and Einstein's gravitational aether, a realm where the concepts of motion, therefore extension, are not applicable. Motion and space come with spacetime, not before.

What we call free space or absolute space, complies with Einstein's aether. This aether is similar to Newton's absolute space but with physical properties. It is where the laws of electromagnetism -- ratios like permittivity and permeability -- are determined. It is not bound by the rules of spacetime and exists independently from spacetime. It is before spacetime, just as your absolutely free of EMR and establishment accepted, free space.

Newton's absolute space, an empty plenum, a void which he considered to be real, was the seed to Einstein's aether. But he saw that, for Newton's view to be correct, objects would have to be perceived as really separated, related only by their macroscopic mechanical interactions (Billiard Ball model). In Newton's view there would be no need for the principles of equivalency and relativity to accurately describe this universe, but scientific facts say we need relativation.

In reality, time and space are related. Einstein's geometrical principles sprang from the same concept of unity and wholeness implied in most of Mach's ideas. In Mach's view there is reciprocity between space and matter. This unity and oneness which distills from the GTR, is the philosophical basis for Einstein's holistic views.

Without an aether there can be no spacetime because there can be no continuity, hence no causality, nor process. This is also what Mach had realized. Resistance to motion, he thought, could only be the product of unity among all objects. He believed that there is a constant dependence (reciprocity) between matter and the space that surrounds it, and that this unity could only be explained by an all pervading aether. But in those days, most scientists preferred to call it by names like absolute space, free space, inertial space, momentum space, the gravitational ether, or the quantum vacuum. All of which refer to the same empty space. Now, they are either calling it the Higgs field or Dark Energy.

Spacetime

In spacetime, there is no such thing as a *perfectly flat vacuum*. A few decades ago, as scientists were trying to measure the temperature of space, they found that it was noisy and full of radiation, or quanta. Scientists then, started to call space by names like cosmic background radiation (CBR) and zero point energy (ZPE), which is akin to John A. Wheeler's pre-geometry, or Quantum Foam. Later, they discovered that throughout the observable universe, the CBR's minimum temperature is almost uniformly set at about 2.7 K (Kelvin). Meaning that it probably is isotropic and homogeneous, that the universe looks and behaves the same everywhere.

Spacetime is a medium with mechanical properties (pressure and density). It is observable, hence measurable, and unlike Einstein's aether, possesses material, or mechanical properties. Space is matter, the aether is not.

Spacetime is made from many different types of particles. Some resist compression, or exhibit negative-gravitation, and some are infinitely compressible, and exhibit positive-gravitation.

Boyer, et al., described the ZPE as fundamental to spacetime, and thermal radiation as a product generated by the motion of ZPE particles, which in turn were buffeted back into motion by this thermal radiation which they themselves had produced. Thus providing the basis for a perpetual motion system and solving the riddle of the apparently infinite energy coming from empty space.

Now, if space is made from particles, then it may be subject to changes in pressure and density, like a gas. Therefore if space particles carried by matter waves continuously condense into material objects, that would mean that the closer you get to the object, the denser the space would be as a function of the object's mass and radius, explaining why gravitic pressure in space flow theories still obeys the inverse square law.

From Relativity we get that a physical system accelerated through space has the same equilibrium properties as an unaccelerated system immersed in a gravitational field. And as Timothy Boyer showed, a physical system accelerated through space has the same equilibrium properties as an unaccelerated system immersed in thermal radiation at a temperature above absolute zero. And at a temperature of absolute zero, a harmonic oscillator in a resting frame of reference, or moving at a constant velocity, is subject only to zero-point oscillations. In an accelerated inertial frame the oscillator responds as if it were at a temperature greater than zero.

As an object accelerates through space, there is an EMR exchange between the object and space as a mechanism in Nature there to maintain thermal equilibrium between the accelerating object and the space that surrounds it. Nature, in order to maintain thermal equilibrium, uses this mechanism, now known as Timothy Boyer's thermal equilibrium radiation, or Unruh-Davies radiation.

When an object is just sitting, but immersed in a gravitational field, what's being accelerated in relation to the object is its surrounding space.

And this is how gravitation is explained: just as photons can be moved by electromagnetic forces, space particles are carried by matter selective, inwardly flowing, quantum matter waves (Gravitons?) in a gravitational current. Space flow manifesting as gravitation. It is material space flowing into matter, and this flow is caused by the force of gravity. Gravitation as described by the inverse square law, because space becomes denser and pressure goes up as space matter gets radially closer to the source of the gravitational force.

Aether and Information

Plato was wrong in believing that there is a separate realm for Ideas and Forms to be stored at.

According to today's physics, as the first fields appeared, information also appeared. There is no meaningful information below the Planck scale. The aether is where the strong and weak nuclear forces, and the magnetic, electric, and gravitational fields emanate from as autopoiesis occurs, in spacetime.

In this view, only some of the information about the present environment, about the present inertial state, comes from the aether as it rules the whole universe, all at once, with just a few fundamental laws (constants, or hidden dimensions... passive information). Thanks to this oneness quality of the aether, the inertial state can be registered instantaneously, allowing all kinds of emergent geometrical forms to observe themselves in wholeness and be self-reflective. Every object existing as a solid body is coupled to a matter wave. Information about the object's surrounding environment being picked and brought in by each object's particular pilot wave. Their displacement through space being accurately and effectively described by mathematical wave functions.

But how could information contained as all kinds of geometrical relationships, that are evident or real only in spacetime, be formed, stored, and accessed by process at a sub-quantum level? It is not, information is always material, and can only exist above the Planck scale. Information is mostly transferred by matter as either sound waves, light waves (colors, shapes, depth), and other forms of vibration. Existing as EMR and matter waves, information can be quantized, formed into particles or wave packets, while transferred unaltered through vast spaces at almost the speed of light.

When we think about the infinite capacity of EM waves to store and propagate information (Television, cell phones, radio, and all forms of EMR), we have to ask ourselves why would we want to go sub-quantum when quanta and hyperspace provide us with such a rich structure? In hyperspace, different objects can occupy the same space, this is why EM waves have a virtually infinite capacity for information storage.

Everything is connected by light. There are matter waves and EMR between you and everything around you, flowing from and into everything, including yourself. EMR bouncing off everything, carrying information regarding each surrounding object's objective state, as matter waves bring needed information about the environment.

The medium in which any given self-organized system develops is full of information which the system needs, and will use, in order to maintain itself. Matter is not different, it can be defined as self-organized, and the medium in which it develops as full of information in the form of EMR. Any given neutron will need to extract from hyperspace only the information neutrons need to support their structure as time goes by. That is, neutrons will mostly use information that is meaningful only to neutrons.

During its existence, a particle retains more information

about the past (a history) than it has about the future, creating an information potential (a pressure with a direction), thus the emergence of a quantum potential. From this potential we can partly derive the particle's future. (Yakir Aharonov's teleological state vector, which he probably derived from David Bohm's equations.)

For any process to continue to evolve as a whole, there must be a function of internal oversight. And that is what we have in matter, a self-reflective mechanism in the form of a spherical standing wave, which is only possible when all the parts are interconnected. We know this needs to be a non-local function, and this function is only possible because of that oneness quality coming from the aether scale. It works like a hologram, where each point knows where all the other points are, and where state is instantaneously reported throughout the system. Thanks to this oneness, all kinds of emergent geometrical forms are observed in wholeness as a self-reference mechanism inherent to all matter.

It was from analyzing this holistic awareness function of Nature that David Bohm conceived his quantum potential (Q) concept. At the same time he explained EPR type phenomena (entanglement, non-locality, FTL communication), this self-reference function was defined by his quantum potential concept as a holistic awareness function of matter. The aether being what allows EPR type phenomena to take place.

EMR was, and still is, the most used means of information propagation and natural communication within and between spatially separated objects.

Already existing information leads to the creation of more information. And this information can only be stored as matter, it can only exist in material form, e.g., DNA.

Logos and Nous

Zeno and Democritus were correct, there is a non-material aspect to reality that is not subject to change. That should have been the ultimate truth. The aether has no means for storing meaningful information, but it contains the ratios... the constants... the law. Ratios... real but non-material quantities... the tools of logic... the gears of physical reality.

Logos is the driving principle behind Nous, the controlling principle in the universe. It is before Creation. Logic (Logos) is before knowledge (Nous). It is the passive information embedded in empty space, or in the aether. And all matter, or active information, follows this passive information. Logos represents the ratios, the physical constants (logoi), and as many ancient philosophers already claimed, it is not subject to change the way matter is.

Stephen Hawking once said, matter swallowed by Black Holes is lost information. So, if this were true and this were a cyclical universe, all information would then be lost with each contraction, or crunch... but not Logos. (Hawking later changed his mind.)

When we ask: what came first, Mind or matter? We must first clarify what do we mean by Mind. If Mind represents logic, we can then say that Mind came first, but if we take Mind to be synonymous to intelligence, or knowledge, then Mind came after. See, in terms of philosophy, we have a problem with semantics. The term Mind can lead to confusion -- many confuse it with human thought, or consciousness. We should continue to use the original terms, namely: Logos and Nous. Logos for logic, and Nous for intelligence, or knowledge. Logos and Nous, instead of Mind and matter. Why change the terms our ancestors were already using?

I agree with Heraclitus, materially speaking, the aether (Logos) remains unchanged. The aether acts as a traffic light directing energy flows, which always does exactly the same expected things. It simply reacts to forces coming from matter, or Nous, as Hegel and Anaxagoras called it. In their view, Nous is the universe and it evolves as it learns.

Aether (Logos) only holds the law, what becomes is matter itself. The aether cannot learn, but it can be in different states. It thinks through matter, thanks to matter, but once that matter is gone, it returns to a singularity... or to being just aether, but in a different state.

Quantization and organization of particles in hyperspace is determined by the exclusive dimensions (information) of a myriad of interacting matter fields, which originate from already existing matter, dead or alive. This means, there is no information being projected into spacetime from Plato's Ideas realm, the information comes from an already existing reality, at the spacetime level.

I believe there is Logos (aether), then there is Nous (the universe). Logos being passive information, and Nous, active information. [Brahman and Prana in some Eastern philosophies.]

Most of the information about a material system is contained within the system, but not all. Passive information, which is mostly related to the momentum and location of the particle in that part of the world, is instantaneously registered throughout the system. There is information constantly flowing in and out of every wave-particle system, as self-reference and autopoiesis take place. That is where Cramer's Transactional Interpretation and Wheeler's Delayed Choice, or back-action come from. But for the metric to have these wholeness properties there must be a unity, and this unity can only come from the aether's oneness. Information has a speed limit, but not state. Information can be transmitted by material mediums and then stored as matter in spacetime, but thanks to the aether, state is instantaneously registered throughout the system.

In a computer, you need a CPU, a hard drive, and some RAM to make it work. The CPU is the place from where the flow of information is controlled, but the information itself is contained in a hard drive. In this world, the aether is the place from where the flow of information is controlled, while the information itself is contained in matter, as matter.

Compared to our brains, RAM in a computer is where

each program processes its particular information, while our brain is where we process our individual information. We have the human brain then serving as a natural, state of the art interface between the aether and every living human being, just as there is an interface between the CPU and every program living inside a computer. In other words, we are the programs inside the computer... or the universe.

Process and Continuity

"Reality is not about things, it is about process" --- Lee Smolin

Process happens because of the rules of Thermodynamics and the motion caused by the electromagnetic properties of the aether. It was from this motion that order arose as random events started to replicate themselves, giving rise to cause, time linearity, and meaning to their movements.

Before there can be any meaningful process or causality, there must definitely exist continuity. Because the aether is one, it brings unity and continuity of process. Its main function is to support spacetime as a unified whole, as a continuum, providing wholeness and interconnectedness.

The aether is motionless, therefore changeless. It has no parts, and there is no process within it. Most of the process occurs at the hyperspace and spacetime levels. Motion, time, extension, size, a beginning with an ending, are all non-applicable notions, and yet everything, from quanta, to space, to galaxies, emerge from it. It is nothing in particular, but has the potential to become anything.

If you stop an atom from spinning, it disintegrates. Materially speaking, all there is, is process. Anaxagoras was right, just as Hegel, B. Russell, Whitehead, and Einstein were. They saw that continuity is as essential to process as discreteness and particulation are. Logically, this continuity, this unity, can only be provided by an all pervading aether.

Heraclitus maintained that everything is in a continuous state of flux, or change, but Parmenides rejected this view by saying that, for there to be change, a substance would have to transit through nothingness in order to become something else. He thought that the basic substance must never undergo change. Process, in Parmenides' view, was incompatible with being, but there is process because there is aether, even though the aether itself remains unchanged. Now, Basil Hiley says -- *being is a relative invariant in the overall process of becoming.*

In hyperspace, there is mostly disordered motion, there is no significant order nor process coming from it. Therefore it can be interpreted as discontinuous and acausal. Motion with no time-line to follow... until ordered by fields emanating from already existing, self-organized matter. Spacetime, on the other hand, as the size scale gets larger and indeterminism fades out, follows the rules of determinism and causality.

Objective reality is a continuous, self-maintained, thermodynamically open process developing in a sea of discontinuities (Wheeler's Quantum Foam). Atoms are open thermodynamical systems, there is always energy/information being exchanged between matter and the environment. The same can be said about biological systems: they are open energy dissipating systems restricted by the laws of Thermodynamics.

There is a physical relationship between the substrate at a sub-quantum level and the objective universe. David Bohm called it Quantum Potential (Q). The concept of potential, in both cases, electrodynamics and Bohmian mechanics, represents a non-local connection between the non-dimensional aether and spacetime. In Bohm's view, mind and matter draw information and energy from their common background through a process of quantum decoherence where Nature can take a look, through wave superposition, at a number of possible outcomes, all at once, as it chooses what it needs. Emerging from all this, we get wholeness, non-locality, non-linear parallel information processing, and very high degrees of order and complexity.

Take a look at waves as an example of non-linear process and you will see that, thanks to wave superposition, you may have one simple wave which could be contained as a whole within a more complex wave, which may in turn be contained as a whole within an even more complex wave... to infinite complexity (Fourier). Because in hyperspace more than one object can occupy the same space, and because of the aether's oneness, activity within hyperspace does not depend on linear time; wave superposition can be a non-linear, parallel information processing mechanism.

A particle maintains its shape and internal structure, as it moves through space, by keeping intact the geometrical relationships that exist between its parts and the environment. As process moves forward, particles must continuously reset their geometrical structure, having to continuously process all kinds of already existing and relevant information in order to continue their existence. They are in fact information... active information. Thus, there is no significantly ordered information contained in hyperspace, just randomly fluctuating quanta which is ordered as wave-particle systems move through it. Space is full of energy in complete and utter disorder, and fields are simply ordered space. Order, or information, is what allows particles to form as 3D space.

Reality is the process from where the objective universe emerges, but the real, the eternal, is the substrate. As a particle moves through space, it does not displace it, it ordinates it. This means that what is really moving through space is an organizing process; matter as active information trying to maintain the geometrical relationships between its structural parts.

As you move, that which was a second ago is no longer,

it is gone forever. No particle can remain floating in space unchanged and immutable. No process means no objective reality. For any object to continue to exist it must always be internally changing and adjusting to each new position in space. If you were to stop its internal motion, or process, it would disintegrate into quantum matter bits (Q-bits) in a very wide radiation spectrum.

The information contained in a particle's pilot wave (morphic field) is what particles need and use in order to continue being what they are. For example, a neutron uses only the information it needs for it to continue being a neutron. Particles, through a process of wave superposition and quantum decoherence, receive, select, and integrate into their wave packet only that information which is important, or meaningful to them.

Reality as Process

Matter is a continuous, time dependent, self-organizing process. Particle-wave systems, as they move through Wheeler's Quantum Foam, will continuously re-ordinate the space that constitutes them. These systems must be in constant motion, continuously processing space in order to continue their existence. If motion were to cease, they would simply disintegrate.

You could compare the movement of matter through the aether to the movement of waves in a pond in the sense that, what is really moving is the form, the information that gives its particular shape to the waves. Reality is like a wave propagating in water... the form moves, carrying information, as water molecules are left behind. As we displace ourselves, it is the information that constitutes our material body which is really moving, the underlying medium stays virtually unchanged. The particle is more like a pattern of motion than a solid, permanent, and space independent thing. Picture the rings that form and spread outwards after you throw a rock in the water, now hit rewind and picture the rings moving back, inwards, towards the rock. The particle represents actuality, as the inwardly flowing waves bring information about the past, about what already happened to the particles around it. The rings going towards and outwards the rock being, in reality, spherical. It is a self-reflexive spherical standing wave that is constantly feeding the particle information about its inertial frame and environment, as a trajectory develops in relation to the surrounding space.

Inside Matter (crystal lattice)
The black dots now seen as the energy source, and the space between them as Einstein's spacetime.

Order arose from interactions within chaos, giving rise to self-dependent, separated cybernetic systems, which in turn evolved to take full advantage of matter's properties. After billions of years interacting with each other, just so Nature could continue to improve its methods to use information to its advantage as a tool against entropy and disorder, these systems evolved into human brains for the purpose of experiencing/perceiving/measuring existence from a 4D perspective. Why? Because all thermodynamic systems tend to move towards thermal equilibrium, and in doing so, they interact, forming physical geometrical relationships in the process, which are then preserved in matter and are accessible by these cybernetic systems, when needed, as they evolve towards stability and thermal efficiency.

Hyperspace is filled with information bits, matter precursors, a pre-geometry comprised by material units of information which exist in chaos and are ordered by logic and activity into spacetime. But for natural reasons, i.e., energy conservation laws, everything that comes into spacetime must be registered and energetically measured before it can materialize.

As the quantum matter that constitutes it flows radially from hyperspace towards its center in spacetime, there has to be a quantum measuring process used to register the particle's location and momentum in relation to the rest of that inertial frame before it can crystallize. Thanks to wholeness, each particle senses the other and their relation to space, building an information network filled with geometrical relationships (spacetime), which are in turn used as the future is built on the already existing information.

Information emerges from and is preserved as matter (e.g., DNA). The components, the ones generating this information, are all floating in and interacting with space as particles, molecules, galaxies... brains... all of which continuously exchange information as they continuously emit and absorb electromagnetic radiation. EMR is one of the tools which Nature has been successfully using since the beginning of time to overcome space-like separation between objects in order to evolve as a whole.

The aether helps maintain the relations that give shape to a particle as it moves, but that information stays always bounded by the laws of spacetime. For example, virtual particles that pop-up from hyperspace into spacetime are the result of random field interactions such as resonance. Particles such as fermions stay in spacetime because they have stable standing waves. So, unless you are a theoretical physicist, don't be too bothered by all those subatomic particles being detected, only the ones we find in our every day lives have real significance to us.

The aether contains all the energy available to the universe. As bodies move through empty space, stress-energy tensors around that body must be automatically adjusted according to the existing environmental conditions at each point in space and time. And this is possible because of the continuous nature of the aether. That is the principle behind inertia and momentum. The time a moving body takes to reach thermodynamic equilibrium with its surroundings, is directly proportional to the inertia and momentum forces it will experience.

Each body's energy requirements is tied to the total energy available to each given point in space and time.

As objects are accelerated in a space-time metric, each new position creates new energy requirements from the continuum. The aether provides the unity needed for the inertial state to be instantaneously registered.

Particles follow the laws of spacetime. Laws by which their structure is going to be held intact in spacetime, only if the spatial relationships are maintained, in spacetime.

If we allow for three different size scales of reality, with aether as the substrate, the whole process can be coherently put together. We could say that aether, hyperspace, and spacetime are all different states of the same entity, the difference between them being just a matter of scale. At the aether scale, there are no reference points. No meaningful motion, no time, and no wave fronts, with empty space as perfectly flat. Then, at a larger scale, it becomes hyperspace, and we get things like EM fields, photons, bosons, and all types of quanta. Finally, at an even larger scale, we get Einstein's spacetime, the objective universe as we naturally perceive it.

> 1. **Aether** -- (Sub-quantum level)
> Dimensionless, timeless. David Bohm's implicate order. Basil Hiley's pre-space. The scale at which non-local, instantaneous state change occurs. This is where Nature's four fundamental forces are administered from.

2. **Hyperspace** -- (Quantum level) Wheeler's quantum foam. Quantum gravity. Quantum weirdness. A medium where more than one object can occupy the same space; allowing for quantum wave superposition and parallel, non-linear information processing.

3. **Spacetime** -- (Classical level) Inflation, Gravitation, Black Holes, matter... the objective universe.

We could conceive reality to have started with this non-material, indivisible substance from which hyperspace emerged. A reality that shares the properties of both: the aether's omnipresence, and hyperspace's non-linear information processing. Which then turns into spacetime as reference points and linear time emerge, sharing the properties of all three scales, or realms at once.

According to Timothy Boyer, Quantum Foam (space) is constituted by at least two different types of quantum matter. One is noisy and expanding, while the other is ordered and condensing. One exhibits negative gravitation, the other, positive gravitation. From one, space is created, from the other, matter is created. Quantum matter waves (Q-bits) inwardly flowing into matter (Gravitons?), as heat and light continuously flow away from matter. There is a continuous condensation and expansion of space taking place.

We need two separate matter states, two types of motion

with opposite directions, one deterministic, the other indeterministic. Objective and subjective. Matter and quantum matter. Particle-wave systems in a continuous exchange of information between matter and space in a non-ending quest for self-consistency. Order (matter) coming out of entropy (quantum matter). It is a semiclassical process, an endless information exchange between the metric and quantum matter. Schrödinger waves evolving in curved spacetime.

The Schrödinger equation must be solved according to the spacetime geometry, but then, as the system continues to move in spacetime, stress-energy expectation values need to reset to a new spacetime geometry, making it necessary to again solve for the new quantum state. This process repeats itself in a continuous cycle without end in the form of spherical standing waves.

It is a process where an object's matter waves are described as being continuously condensed by gravity.

Leon Rosenfeld (1933, 1963 papers) considered:

Where (psi) represents the quantum state of matter fields, and:

> * Gravitation is the beginning of a semi-deterministic process. The gravitational field (G) is considered a classical, non-quantized

field.

* Gravity and wave packet collapse are interrelated. There is wave packet collapse because there is gravity, gravitation being the product of gravitational forces generated within matter.

* Gravitation is caused by a space pressure differential created by the inward, radial space flow that occurs as a result of the object's wave packet continuous collapsations.

* Quantization and organization of space is orchestrated by matter fields which originate from, and follow, exclusive dimensions already existing as matter. Energy being quantized into particles by spontaneously emitted sub-atomic particles (Higgs boson?), in hyperspace.

If there were no wave packet collapse, there would be no matter, simple as that. Stopping or reversing gravitation would require the stopping or reversing of the matter wave flow, and that would cause matter to disintegrate. All matter is wave and particle at the same time (de Broglie, et al.), that is why there is gravitation. So, let's stop wasting time on anti-gravity devices and let's concentrate on creating a spacetime bubble in a completely independent and separate matrix. A bubble which could be extracted from and reintroduced back into the time-cone at will.

The Universe as a Hologram (my interpretation)

Gravity as a negentropic force? As an information gathering mechanism? That's what it looks like.

Let's look at a our galaxy, then apply this model to a subatomic particle.

At the center of our galaxy we have a black hole, or a singularity. This black hole is constantly pulling matter/information, but all that information stays on the surface (Event Horizon), the black hole's surface growing directly proportional to the volume of the bodies it swallows (Jacob D Bekenstein, Gerard 't Hooft, Leonard Susskind, Juan Maldacena, Stephen Hawking, et al.). So, black holes inside galaxies, like the black holes inside subatomic particles, are basically nothing more than information gathering mechanisms.

All of these black holes acting as information nodes forming a quantum network or hologram (spacetime) where the holographic plate is the two dimensional surface of the event horizon and the non-dimensional object (singularity) in the center of each body acting as their energy source.

Right, all that missing mass (aka., Dark Matter) is now being considered by contemporary physics to be contained by empty space itself, probably in the form of infinitesimally small black holes in the center of neutrons, protons, electrons, and the rest of all subatomic particles. Which is how all matter is connected to the whole. Current physics' description of black holes, singularities and the gravitational aether being actually very similar.

Octopus Wormhole
Olena Shmahalo/Quanta Magazine

It is a radically holistic view of reality where entanglement is seen as the glue that keeps the universe from atomizing. Entanglement made possible by the aether's oneness.

About Gravity, Inertia, and Mass

Today, space is not the same thing 19th and early 20th century physicists viewed as space. Back then, there were no CBR, nor Wheeler's quantum foam. Now, space is considered to be material, a collection of small particles. Space is now chock full of particles, most of which are photons and zero point radiation. This is why modern physics now say space is grainy.

Cosmic background radiation and other particles fill the observable universe. What I call aether is before this material space, it is what Einstein called, the gravitational aether. The gravitational field (aka., aether), as described by Einstein, being continuous, not quantized, as Wheeler's Quantum Foam is. Reality is built on quantized structures. Matter is always quantized but sits on the gravitational field, which is continuous.

In this view, particles in spacetime are gathered and organized by matter waves (morphic fields), as matter crystallizes, or condenses into atoms, molecules, cells... etc. Quantum matter (Q-bits) continuously flowing into matter as it is quantized and carried by concentric, spherical standing waves. EMR radiating from and into matter in a constant exchange of information about its objective state and its relation to surrounding objects in spacetime.

First there is the quantum vacuum, which is made of virtual particles billions of times smaller than any neutron or proton. Tiny particles which are not spinning, but pulsating in and out of spacetime. Then come the bigger and more stable particles, which are made out of the space foam they are continuously feeding from. This inwardly flow of space foam being what causes gravitation, spin and all those vortices (Penrose's spin networks) many are trying to understand. Spacetime formed by a tensor network, with gravitational waves being the result of tensor fluctuations... that's the current view.

Tensor Network
Compared to the previously shown 3D image of a crystal lattice, in this 2D representation the filaments connecting the entangled particles are now seen as the same strings <u>String Theory talks about</u>. In Quantum Gravity Theory these are called tensors. In my view, these are information highways.

Gravity

According to the theory, gravitation started the moment the Big Bang occurred, with gravity acting as a negentropic force always trying to put spacetime back into a singularity.

First of all, let gravity, gravitation, and gravitational waves be three different things. Gravity being a fundamental force, while gravitation and gravitational waves just the products of that force. The way I see it, gravity starts at the aether level as stress-energy tensors and lines of force are formed. Space particles (quantum matter, or Q-bits) being carried by matter-selective, inwardly flowing quanta (Gravitons?) in a gravitational current, the same way electrons are carried by an electromotive force. The center of each particle acting as a sink, or miniature black hole. Wave packet collapse being closely related to gravity (Quantum Gravity).

Imagine the first particle as a point-like object pulsating at a very high frequency in an endless quest for thermal efficiency and equilibrium. Each time it pulsates forming a wave-front pushing outwards, driven out by the force of anti-gravity (aka., Dark Energy), which then contracts as the wave is pulled back in by the force of gravity. This pulsation, or vibration, causing this space (quantum foam) inwardly flow we know as gravitation. This motion seen as an information gathering mechanism in which the wave front is where the information is being stored at.

You can visualize the fundamental particle as a bubble that inflates and deflates as it pops in and out of nothingness (AKA., Einstein's gravitational aether). Which is how most visualize Wheeler's Quantum Foam: as a bubbling soup, the bubbles being virtual particles, and the soup the gravitational field. Bubbles made of energy. Remember, energy is space in motion. Space being stretched, which is where the stress-energy tensor comes from. As the bubble inflates, space is being stretched out, until gravity wins and collapses it. While the bubble inflates, space and time are being created. Before the bubble, there is neither space, nor time, all there is is aether, which is pure energy. Energy which is neither hot, nor bright... until the bubbling starts.

Gravitation coming from a pressure differential in material space caused by the constant radial flow of matter waves into bodies with mass, as quantum matter condenses and crystallizes into its objective state.

Gravitational waves being just the ripples (tensor fluctuations) being caused by the motion of the bodies floating in this sea of particles.

Picture two bodies, like the Earth and the Moon, now imagine space flowing into each body at the same time: that causes gravitation. Because there is space flowing in opposite directions, which causes a decrease in material space density, there is a drop in pressure that makes both objects drift toward each other. That is also how we get tide movement: the Moon casts a shadow as it blocks space flow to the Earth, causing gravitic pressure to drop between the two bodies and consequently causing the sea level to rise where the shadow is being cast.

Gravitation being caused by radial space/information flow, as photons in hyperspace are converted into matter by an autopoietic process driven by logic and the laws of Thermodynamics. Massless bosons, full of information, being used to support the structure of already existing matter, in spacetime. Space being viewed as q-bits of information (J.D. Bekenstein, Stephen Wolfram, Lee Smolin, Gerard 't Hooft, Leonard Susskind, et al.).

Speed of Gravity

Some wonder if gravity travels and at what speed, but the truth is that gravity does not move, gravitons do. Claiming that gravity moves is like claiming that the electromagnetic field moves, when in reality, what moves are photons. Same with gravity, what moves are gravitons. Gravity (the gravitational field) has no need to move, it is everywhere.

Dark Energy

Today, Dark Energy is seen as anti-gravity, or gravity that pushes spacetime out, instead of pulling it back in. Why now? Because of a change in the energy density of the universe, change caused by all the condensation of space that has transcurred to this date. An accelerated expansion seen as a result of too much condensation, or too much matter and too many black holes in the same vicinity.

Inertia

Inertia, as conceived by Galileo, meant a resistance to change, and process is synonymous to change. Reality is process, and inertia is a product of this process.

The force of inertia, or momentum, that force we feel pushing as we come to a traffic stop, comes from the drag produced as the energy parameters between space, a moving mass, and the surrounding masses synchronize. There is a time delay as matter reaches thermal equilibrium with its surroundings, and we experience it as inertia. This is the nature of Timothy Boyer's thermal equilibrium spectrum. It takes time for any moving object to reach thermal equilibrium with its surroundings, as it jumps from inertial frame to inertial frame, in spacetime. Each inertial frame being, in reality, marked by the collapse of each matter wave front, like a quantum clock.

Inwardly flowing matter waves bring some of the information necessary for each quantization before reaching each successive inertial frame's objective state. Each object having its own rate, its own wavelength, or its own pulse, so to speak. Therefore inertia may be viewed as the drag caused by this radial space flow process and the time it takes to occur.

Inertia is the result of aether drag: a change in space flow rate, into and from the particle, as the particle moves through space. It comes from the tension created by changes in space flow rate as a material system accelerates within the chaotic medium.

As required by the equivalency and relativity principles, there is a momentum/information exchange between particles and space, through EMR, as they continuously reset their spatial relationships. (Unruh-Davies radiation and Timothy Boyer's thermal equilibrium spectrum.) When a particle is moving at a constant rate there is no information lag being created by the space/information flow occurring within the particle. Its spatial, energetic, and geometrical relationships remain constant, but as the particle is accelerated, the energetic and geometrical relationships between the particle and space will be in a constant state of change, causing this space flow tension we call inertia.

While a particle is moving at a constant speed and all the geometrical parameters are set, it will not experience any inertial forces, but as it accelerates and the relationships change, it needs to keep re-adjusting to its new energy/space consumption settings. This is why relativistic effects are so real. When accelerated in relation to other particles, space shrinks, time slows down, and mass (process) increases within the particle. This happens in order for the particle to balance its energy usage in momentum space, and to maintain its relationship to spacetime in accordance to energy conservation laws.

To us, at the Classical level, it seems as if it became a bunch of unrelated separated entities, but in reality it is all connected through the all pervading aether. Even the most desolated regions of the universe are part of the one single process that started it all. Simply because in reality there is only one process, a universe from where a myriad of interdependent information nodes (objects) formed to become apparently separated systems, everything that moves will experience inertial forces.

Mass

Mass is caused by the continuous internal process occurring within matter, as space/information radially flows towards the center of all matter.

Information about a material system must be contained within the system, it does not come from anywhere else in space. The only external information being brought to the system by EM waves is about the momentum and location of the particles in relation to the world. These systems are comprised by particles and their particular standing matter waves. Particles being selected, quantized, and turned into parts by standing matter waves as space particles condense into the particle/system. The parts, not the information, to construct and maintain the system intact as it moves through the medium come from the chaotic hyperspace. Wheeler's Quantum Foam is the material space from where, through a process of wave superposition and quantum decoherence, new geometrical information is created and incorporated into the system as it moves through spacetime.

The way I see it, the so-called Higgs field is just another name for Einstein's gravitational aether. Mass being the result of matter's field interactions within itself and the space in which it sits, hence the Higgs mechanism. Matter is a continuous, time dependent, and thermodynamically open, self-organizing process. Particles, as they move through Wheeler's Quantum Foam, need to continuously re-ordinate the space that constitutes them. They are in constant motion, continuously processing space/information. Matter is formed by this process, and mass increases directly proportional to the amount of process. This is why the denser a particle is, or the faster it moves in relation to other objects, the more massive it becomes. Mass is directly proportional to process.

Holistic Awareness

According to present day theory, the total energy available to the universe was pre-set at the moment of its emergence, and the force of inertia tells us that each of its parts must register how much energy is being used in relation to the whole. Each object that moves in space must follow the laws of energy conservation. But how else could the universe register how much energy was being used by an object moving at nearly the speed of light if it wasn't through momentum space (aka., aether)?

As Ernst Mach explained inertia, he came to the conclusion that energy usage by objects within the universe is instantaneously registered through momentum space. This is where phenomena like inertia come from. Particles sense other particles as they complete the state information exchange and realize the spatial relationships required to collapse the wave packet from hyperspace, as they crystallize into spacetime. Holistic perception is an intrinsic function of matter explained by the aether's oneness.

The observer, in the present theory, must refer to any object that is able to interpret environmental information brought in by EM waves. Observation with the only purpose of establishing the particle's spatial parameters (speed, distance, mass...) in any given inertial frame. The particle must first be perfectly synced with the environment before it can exist as matter in spacetime, that is the law.

John Von Neumann was right when he said that the evolution of the Schrödinger wave depends on quantum mechanical observables, implying that this information can only come from spacetime. Yet since the theory considers brains to be quantum measuring devices, it also includes human observers as efficacious agents. The only reason human brains entered the equation was that, as they received light (EMR) coming from the particle, just as all objects in spacetime do, information about momentum and location of the particle, which is vital to maintain energy conservation laws, became known to the particle/system; allowing it to complete the feedback control loop and continue to condense.

So Quantum Mechanics' big mystery was: why do I have to observe Schrödinger's cat in order for it to live or die? The answer is that our brains are quantum measuring devices, just as the rest of all matter is. We are the best quantum measuring device that ever emerged from all the information processing that has transcurred in our neighborhood to this date. Interactions within a system, like in a brain for example, depend on more than the information it gets through the senses.

Perception is a very old natural function inherent to all matter, not some exclusive human ability. In their endless quest for thermal efficiency and equilibrium, particles in spacetime perceive, select, and integrate into their wave function only that information which is important or useful to them. Our mind, with all of its mental waves and accompanying frequencies, became the modern version of that same holistic awareness function after 14 billion years of information processing, autopoiesis, and evolution.

Wholeness in space and time is what allowed Nature to

evolve.

Consciousness

What role does light (EMR) play in determining Schrödinger's cat's state (dead or alive)? Is light itself the only important factor closing the loop, or is the observer's conscious acknowledgment which causes the final determination of the cat's fate? In other words, is the wave packet collapse a function defined by the structures of matter, a result from the interactions and relationships of its parts, independent from human observers, or is objective reduction a function of the human mind? The answer is yes to both questions, there is no contradiction, self-observation is a function intrinsic to all self-organized systems.

Perception is key to the crystallization of 3D reality. Every particle and object is accompanied by a wave that informs it about its shape and exterior environment. The particle exists in 3D only during actuality, at the now moment. EMR feeds molecules with information about the environment (information that represents the molecule's past, as it exists in a point-like 3D actuality). In a spacetime continuum, solidity and volume manifest only at present, or actuality. There is no material past, nor future.

Light particles transcend time and space. Photons bring us the past. Because traveling at the speed of light causes time to virtually stop, information from the past is locked into photons. This is how we can see what the universe was like billions of years ago. Our capacity to see is closely related to consciousness and our ability for self-reflection, just as EMR is closely related to state vector reduction and matter's ability for self-reflection.

We are constantly choosing the present out of an infinitude of possibilities through a mechanism of quantum wave superposition. Thoughts are formed very much the same way particles are, and just like particle systems depend on matter waves and wholeness in space and time, so does our mind. Processes forming ideas are very much like the processes that form matter. Mind and matter, both depend on phenomena like wave superposition, non-locality, and parallel information processing. Phenomena which ultimately gives all matter the possibility and the ability to self-organize into ever more energy efficient systems.

Holistic awareness, or self-reference, emerges from an inward necessity which is satisfied as information is chosen from the context in which a system evolves. That is why experience/perception is fundamental in the development of all matter, but especially in intelligent beings: because we need it in order to be able to choose. This is why Nature (self-organized matter) transformed into brains with eyes: to more efficiently carry out this self-reference function. How could matter get organized if it could not observe itself? Matter, in order to evolve, had to communicate in any way naturally possible (e.g., surface vibrations, air vibrations, EM radiation, and non-local communications). Biological organisms evolved to use light to their benefit very slowly. As we already know, it took Nature billions of years (from the Precambrian to the Cambrian era) just to develop eyesight.

Human consciousness evolved from the same holistic awareness property all matter has shown to possess. The evidence suggests that the objective universe was here before human observers, and that wave function collapse is a very old function of matter, which through a self-reference mechanism inherent to all self-animated matter, evolved to what our consciousness is today.

Human consciousness is spacetime dependent, just like matter. No brain equals no consciousness. First, there had to be matter before there could be any brains, and matter is spacetime dependent. Brains emerged from the evolution of information that existed in spacetime. Thus consciousness appears with the emergence of matter, not before. Spacetime is where experience takes place.

There can be no evolution outside of spacetime. Now, after billions of years, this information exchange between matter and the environment in which it evolves has produced ever more complex self-organized systems. Human beings have evolved to take full advantage of this holistic awareness function of Nature, which is what enables us to think outside the grip of time. Allowing us at the same time to remember the past and imagine the future. Thought and self-awareness can then be conceived as the products of that same holistic awareness function through which all matter started self-organizing almost 14 billion years ago. Consciousness comes from the same holistic awareness function found in all matter.

The difference between humans and the rest of the animal kingdom is self awareness. Animals, with the exception of human beings, are bound by time, they exist frame by frame, and react accordingly. Humans, on the other hand, have the ability to voluntarily go back and forth in time, we call it imagination, foresight, or insight, and that is what gives us our sense of wholeness in space and time... which is what human consciousness is all about.

Human consciousness is the ultimate product of a natural, energy balancing mechanism, determined and regulated by the laws of Thermodynamics. Because energy is finite, each object's energetic requirements have to be measured before entering any given spacetime metric. Before going from its subtle quantum matter state, or wave state, to its objective material state, or particle state. As described by Quantum Mechanics, these information requirements are met through wave interactions and the mechanisms governing wave superposition.

Brains are these little bio-mechanical tools that emerged with evolution for the only purpose of enabling us to interpret and interface with reality. In order to become more thermally efficient, the universe needed to improve its abilities to observe and perceive the environment. After billions of years, matter evolved into brains that could take advantage of the properties of spacetime.

Brains exist because there is spacetime, not the other way around. Human sentience is the actual evolutive result of all the sensing matter has been doing through time. Matter is aware of its surroundings, but that does not mean that it can think, not unless it had been previously formed into a brain.

Nature would still be able to exist and observe itself without the human observer, it would just be a more primitive process. Our brains then, are seen as Nature's best developed self-reference tool on this part of the universe. Human consciousness being an extension of the same holistic awareness function self-organized matter always utilized to observe itself. Therefore in a very real sense, human consciousness is still Nature observing itself.

Experience is fundamental to existence, but it is not reality. Berkeley was wrong. Reality is the process through which Nature is constantly becoming. As Sir Roger Penrose explains quasicrystal development in *The Emperor's New Mind* (p. 564), he writes that while constructing their *"randomly forbidden, very complex icosahedral symmetries"*, and using wave superposition as a self-reference mechanism, it appears as if the whole crystal is observing itself, registering all atom configuration patterns embedded into its pilot wave. Their present state being compared to past states and all the possible outcomes, all at once, until the right atom configurations are found as quantum decoherence takes place.

Experience plays an important role in the correct

development of the crystals, as well as in all self-organized systems. The crystals are able to carry out their self-observation by following information contained in their pilot wave (Bohm-de Broglie), which contains past and even future information about the crystal as a whole. Proto-qualia, or state for a quasicrystal, would be like how all the possible atom configurations would feel as they remain in superposition until the right one is found. Then, and only then, could the collapse of the wave packet finally occur. Build a machine which can follow its pilot wave and fully register its quantum state, and we may finally have a self-organized, and maybe even, a self-aware machine.

From the moment the first self-organizing systems appeared in Nature to the moment the first human brain appeared it has been a few billion years, but in both occasions the purpose has been the same: to experience existence. Penrose's quasicrystals do not have a brain, but they follow their morphic matter wave as the measure by which they must exist, and if by any reason they were to stop following it as they added new atoms to their body, they would end up becoming a totally different type of material. The objective state a human being follows, or the measure by which a human being exists, is also defined by its brain wave-function.

Quantum Entanglement

John S. Bell was right. As already confirmed by Quantum Mechanics, the universe violates locality at the quantum level. Local realism applies only at the classical level.

The collapse of the wave packet on the EPR and Aspect experiments doesn't just come from human knowledge acquired during the measuring process, but from a holistic awareness property intrinsic to all matter. And, as Eugene V. Stefanovich contends: interactions, not forces, are instantaneously registered throughout space.

Many are amazed at Wheeler's Delayed Choice experiment results, but that is because they want to understand it from their own perspective. They want to understand it applying spacetime rules, and that is the problem. At the quantum level, you need to toss away the notions of time and distance. For you, who live at the spacetime level, the photon may have traveled billions of miles while taking millions of light years to arrive, but at the quantum level, its emission, detection, and measurement, all happened almost instantaneously. In our world, it appears that, as we *measure* the particle, we are deciding the path the photon had taken millions of years before, when in reality, the emission and detection of the particle happened almost instantaneously. After emitted, the photon remains timelessly suspended in hyperspace until detected, or measured. Then, as its state is decided and the state vector collapses, it materializes into spacetime: our level of existence. So, it is not our knowledge which collapses the state vector, it is perception. What perceives it could be anything, dead or alive, the particle's state will be defined either way.

As we already know, matter and space are one and the same thing (Einstein), matter tells space how to bend and space tells matter where to go (Mach). The way science sees it, matter is nothing more than condensed space. And, in my view, when I talk about a particle, I might as well be talking about a human being. To me, a man is nothing more than an uber particle, so to speak. That is because I believe in the evolution of matter. Matter, in my view, is synonymous to information, active information.

Also, as we should know, there is a wave-particle duality (de Broglie, Schrödinger, Bohm). The particle is always accompanied by a self scanning, standing wave, where most of the information concerning the geometrical properties of the particle is contained. This wave is called a matter wave, and can be mathematically described by a wave function. Each time the wave function collapses or we have a state vector reduction, the point-like state of that particle is defined in spacetime, as required by spacetime laws. This process is known as quantum decoherence, and works through wave superposition. The particle existing in a point-like state only at the now moment. Which is why it vibrates.

At the quantum level, motion occurs in a similar way it is created on a TV screen. As you may know, a TV screen refreshes 60 times per second or so, that is how motion is created. Imagine the fundamental particle as a vibrating 3D system refreshing its structure over a trillion times per second (Planck time). A standing, self-scanning, spherical wave with a point-like particle at the center. [Motion and time may seem like illusions, but the process that is reality is certainly not. Now, quickly rolling a film in front of the lens inside a movie projector in order to produce a *motion picture*... that is an illusion!]

So, each time there is a wave packet collapse, the *now* state of the particle is defined. Then and only then, can the particle materialize into spacetime, where the laws of Thermodynamics and Relativity apply. Without the information required for the wave packet to collapse, there can be no particle in spacetime; it may exist virtually, but not in spacetime.

Unified Consciousness Field

Now, there are the Bose-Einstein condensates: a state of matter where groups of particles exist under the guidance of one single pilot wave, as a whole. Also known as super atoms, or superwave functions, these were Bohm's main concern as he wondered about the relationship between human consciousness, the body, and the Universal Mind: a realm he likened to Plato's realm of Ideas and Forms.

We now have that each particle's existence in spacetime is defined by the information contained within its pilot wave, and that state is instantaneously registered throughout space. Each particle going from state to state, with each new state superseding the previous one. Learning and evolution made possible thanks to this super fast, continuous succession of states.

So, is the brain accompanied by a unifying superwave function (unified consciousness field), the field where Dawkins' memes are to be contained? There is evidence that points to an affirmative answer. As we already know, the human brain is divided into two hemispheres connected by the corpus callosum, the structure through which both hemispheres communicate. Well, there is a procedure for the treatment of some cases of epilepsy where the two halves are surgically separated, and people who have gone through this procedure still think as one single individual. Meaning that consciousness may be such a higher function that it exists as a non-localized field, where this field's features are determined by the brain, like a hologram. Information being processed in a way similar to how a 3D image is reproduced out of a holographic plate. The brain being the holographic plate, and your thoughts, or imagination, being the reproduced images on the hologram, or field (K. Pribram). State being instantaneously related to the whole brain as emotions. [This, I believe, is how meditation can help with the control of some bodily functions, and even healing.]

And there you have it: particles and cells have a self-scanned nucleus and we, as indicated by gamma waves in EEGs, have a self-scanned brain.

Have you heard the Dalai Lama talk about compassion? I like to visualize consciousness as a field around our head, a field with a given circumference. A sociopath's field having a circumference, or radius, of only a few inches. While people who, like the Dalai Lama, are full of compassion, have a field of a much greater radius... as they reach Nirvana... *like a circle whose center is everywhere and circumference nowhere...* (Zeno, Pascal, Bruno).

Thought and Determinism

GTR is an idealization of reality, a method, a mathematician's trick to eliminate all local degrees of freedom (uncertainty). Smooth-out spacetime, and you get theories like Relativity to work. But there is a background (Wheeler's Quantum Foam) without which there would be no material world.

Schrödinger's equation develops in a sea of real uncertainties (background radiation). As reality unfolds, none of its possible outcomes exists prior to the wave packet collapse. As the Schrödinger wave evolves, the system will have some tendencies, or propensities that depend on the system's properties in spacetime. There will always be some preferred outcomes whose probabilities are going to be higher than those which are not as well related to the system.

Particles exist in a field, but you could never tell exactly where, you could never have 100% certainty. Quantum particles are both particle and wave at the same time. As Louis de Broglie figured out, each object, regardless of size, is accompanied by its own particular matter wave. After a particle condenses, it still remains wave and particle, even after observed and measured. Because all objects are wave and particle at the same time, they are always in a state of motion. Even though indeterminism fades away as objects become larger, there is always an amount of uncertainty left which allows for some degree of freedom.

If you do not allow for indeterminism in Nature, then you must also reject Darwin and the evolution of species. What need would a theory of evolution satisfy, what purpose would it have in a world where everything is supposed to be known from the beginning? If there were strict determinism, there would be no deterioration in objects, things would remain intact forever. There would be no ever growing entropy, just as there would be no need for change, process, or becoming. There would be no need for motion, no need for time, and no need for death!

There is teleology determined by Thermodynamics, but nothing is predetermined. Subatomic particles follow an undetermined path, it is the nature of reality. The motions of all objects in spacetime the size of an atom and bigger follow semi-deterministic laws. Aside from the reality of quantum uncertainty and the fact that, as proven by Nobel Prize winner Ilya Prigogine, sub-atomic processes are not time-reversible in self-organized matter, thought itself may also be largely ruled by indeterministic quantum mechanisms.

Everything in our normally perceived reality follows the

rules of motion as described by Classical Physics and Relativity, but our thoughts may follow the laws of quantum indeterminism. We are choosing the future out of the multiple paths it could have taken. The path my thoughts take cannot be predetermined. Thoughts, as well as sub-atomic particles, have the potential to follow several paths. That is the beauty of quantum mechanics. A determination of the path a thought may follow depends on what we focus our attention on. That is, thoughts, like sub-atomic particles, *happen* as we observe and perceive them, and this is what free will is all about.

Quantum mechanical process is indeterminate going forward, or backward. Reality, like thought, is about becoming, it is process, and this process is totally dependent on the uncertainty that exists in the movement of quanta.

Uncertainty is what broke the symmetry, it is what turned flat space into curved space, it is what causes activity. It is due to the natural indeterminism of quanta that the universe exists. Take the uncertainty away and it will stop moving. Uncertainty is intrinsic to the evolutive process. Uncertainty, which at the human scale (spacetime) translates to a question like... *What am I?*

[Personally, I like to visualize this symmetry breaking as a Yin Yang circle (Singularity) whose halves become unstable and out of sync, then explodes (Big Bang), as the two collide with each other.]

Syntropy and Evolution

Evolution is this very slow, ratchet-like motion, where Nature selects and locks-in any advantageously occurring changes. These are property based selections where the favored properties are usually the ones which will lead to increased thermal efficiency, as information works to fight entropy... or waste.

Self-reference is the fundamental property of Nature that made possible biological evolution. This is the same property that has made it possible for matter to naturally evolve into human brains as the ultimate self-observing example of self-organized matter.

The universe exists because of active information contained in all kinds of interacting waves, and if it were not for wave superposition, there would be no universe. Inherent to wave mechanics are the mechanisms of wave superposition and parallel, non-linear information processing. These mechanisms, which are also affected and regulated by the laws of Thermodynamics, are in great part responsible for information growth and the evolution of biological matter.

Perception, sentience, and communication are fundamental functions of all matter, dead or alive. Sentience is not unique to living matter, but living matter is the result of sentience. Sentience is synonymous to communication, and communication is any information exchange between any two or more parties. The observer could be a molecule, a rock, an ant, or a human being. Take neutrons for example, how can a neutron maintain its geometrical configuration during millions of years without breaking apart? You can also call it self-reference.

Objects in spacetime are continuously interacting with space and the environment through EMR and matter waves, always following classical and quantum field laws. Information about the environment and other objects is transferred to objects through EMR and matter waves. And inertial state is transferred through the aether, or as many prefer to say, through momentum space. Wholeness in space and time, unity of process, only being possible because the aether is one, and this unity being what allowed Nature to evolve.

Some researchers specifically linked the so-called *measurement problem* to humans with eyes; as if someone had to be looking before we could have a measurement followed by a collapsed wave function. But, as experienced by many scientists, it is now known that even when there were no human observers involved, there would still be some form of communication between the instruments being used and the observed event. Before there were eyes, there were other forms of communication going on through matter waves and many forms of EMR between atoms, molecules, bacteria... etc. Quantum measuring is a built-in function of all matter.

Using a Leaf Wing butterfly to explain God's intervention

in the creation of its camouflage was a mistake I used to make. Now, I understand that creating a camouflaged appearance to survive is just another function integrated into their pilot wave function, an algorithm, a program in charge of acquiring information about the environment in order to copy the shape of surrounding leaves and ensure the species' survival. This information is gathered by the insect's matter waves, as well as being supplied by all the matter waves (morphogenetic fields) and other EMR coming from the surrounding environment. As fields are superimposed in hyperspace, meaningful information is integrated into the insect's pilot wave.

The term, *morphogenetic fields*, comes from a hypothesis in which all matter, including living matter, is described as being shaped by a very peculiar type of fields called, *morphic fields*. This hypothesis stands on quite solid ground, scientifically speaking, as it is compatible with the de Broglie-Bohm pilot wave concept. According to this new hypothesis, matter is organized by fields in a similar way a VCR would record a TV program into a magnetic tape. As the tape rolls inside the machine, information contained in EM waves is used to re-arrange and organize the tape's magnetic particles, which can later be used to reproduce well ordered images on a TV screen. Morphogenetic fields supposedly re-arrange and order what are otherwise space particles in a chaotic disordered state into organized matter, e.g., chromosome formation during cell division.

Some say entropy is always growing, at the same time they say energy goes from hot to cold as equilibrium sets in. But how can entropy be growing at the same time equilibrium is setting in? How does syntropy come about? Why does information, especially the more it grows, evidently seeks to preserve itself? Look at human DNA, it still contains genetic code segments that were already present in bacteria 3 billion years ago. Information creating information: is this an intrinsic property, or is there a teleological force guiding it? Complexity and information growth obviously go together. Information begets complexity. Is it an algorithm inherent to biological, self-animated matter? Is this property a product of evolution? The reality seems to be that self-organization started with the first atom, that it was present from the beginning.

Is this proof, however, that Nature is guided by some syntropic quality, a force, an energy analogous to information, ruled by logic, which will try at any opportunity to beat entropy by constantly working to create order out of chaos? Who is the designer, is it the universe itself, or is there an external designer or creator? The question is, is there an external guiding source of knowledge, or does the knowledge needed for these gradual changes to occur exist within matter itself? I believe matter follows logic. Process follows logic, but the intelligence, the geometry needed to build the chemical components, the molecules, etc., that comes from the universe itself, not from some external mind or creator.

Being, Will, and Purpose

Information (geometry) starts with the quantum. Existence starts with the quantum. Before the quantum, there is aether. There can be an aether without quanta, but not quanta without an aether. Matter is dependent on the aether (aka., the Higgs field), it depends on the background as an energy supply, hence wave-particle complementarity.

At the beginning of time there was a change in state, a phase transition, and symmetry was broken. We went from equilibrium and order, to instability and chaos. From a singularity, to a universe. From certainty, to uncertainty. Reality went from a simple state, to an ever increasing complexity. From an empty and perfectly flat vacuum state, to an objects full, curved spacetime. From not just being, but also to existing.

According to contemporary science, the universe is ruled by four fundamental forces and logic, the rest happens by chance. Electrons always move the same way, just as magnetic fields always follow the same rules. Those things are always controlled by something else, they cannot choose which way to move. The aether acts like a traffic light which directs energy flow as required by all types of field interactions. There are no decisions being taken, the field will always interact in the same exact manner. Neither the universe, nor Consciousness, or whatever you prefer to call it, cares about us. The day we are gone, the universe will continue to exist as if nothing had happened. Sorry, no Promised Land, nor an afterlife, those are physical impossibilities.

There is no death, only renewal. God is, but cannot exist unless it exists as matter, as a universe. I see the universe as God in its material form, and matter as an instrument to get the information it needs to constantly re-create itself. Considering that matter is made of fields, that if there were no fields there would be no universe, then we could safely assume that all objects that exist as matter are little more than ephemeral images which exist only temporarily. Objects come to be within that which *is*. They *are* thanks to the aether they come from. If you are because of the aether, then your being comes from the aether. You, the biological unit, because of the rules that govern matter, are only for a relatively short period of time.

The universe is where knowledge comes from, and that is what existing is all about... seeing, learning, and becoming. If information were really contained by some universal mind, then what need would Nature have for a DNA molecule? And, if there were an all knowing mind, a creator, why would it need this long to finish its creation? (13.7 billion years and counting)

I think, therefore I am -- said René Descartes. I say -- I think, therefore I exist. I agree with Carl Jung's materialistic interpretation: the collective unconscious as a morphic field where fields are considered to be a form of matter. Teilhard de Chardin was on the right track.

Just as all objects, the Earth has its own superwave

function, which includes each species' particular pilot wave. And just as self-reflection is a built-in feature, gathering and preserving information are also built-in features in matter. That is what all these species have been doing since the beginning of life. We, just as all matter does, also want to acquire and preserve information as we process it (or as we think). As far as we know, we are the repository and gate keepers of all the acquired understanding in this part of the universe.

"An act is a temporal process, and self-inclusion is a spatial relation. The act of self-inclusion is thus "where time becomes space"; for the set of all sets, there can be no more fundamental process...

Every object in spacetime includes the entirety of spacetime as a state-transition syntax according to which its next state is created. This guarantees the mutual consistency of states and the overall unity of the dynamic entity the real universe...

...thus, we can speak of time and space as equivalent to cognition and information with respect to the invariant semantic relation processes, as in "time processes space" and "cognition processes information"...

It follows that the universe freely determines its own constraints, the establishment of nomology and the creation of its physical (observable) content being effectively

*simultaneous and recursive. The incoversive
distribution of this relationship is the basis of
free will, by virtue of which the universe is
freely created by sentient agents existing
within it." --- Christopher Michael Langan
(www.ctmu.org)*

Ponderable matter has its own refresh rate, like a TV
screen, just that it happens at a much faster rate, and in
3D. As described by contemporary Quantum Mechanics,
each particle is accompanied by a continuously
collapsing spherical wave, each full collapsation
representing a moment in time, as the particle jumps from
inertial frame to inertial frame, while it progresses in
spacetime. This is where motion comes from. This is
what makes evolution possible, or how else could state be
preserved if or when information were not? It is a
learning mechanism. Objects, including the universe as a
whole, go from state to state, as each new state
supersedes the previous one. All this information
replication and preservation being caused by this
continuous succession of states. Matter going from state
to state in a direction determined by how balanced, or
coherent, the now state is... or feels. Is this where
emotions come from? Is this qualia?

Meaning

Meaning is intrinsic to sentience. As cybernetic systems
go from state to state, these will keep or reject data
depending on its usefulness. In other words, depending

on its meaning or significance. Meaning being discerned through a mechanism of wave superposition, or parallel information processing, which acts from top to bottom and bottom to top, all at once. Self-replication, learning, memory, self-organization, etc., all depend on process, and for these mechanisms to work, these systems must be able to go from one state to the next. As the wave function collapses, each collapse, each wave-front, representing a new state, each new state representing actuality, or the now moment. The particle being in a point-like state only during this moment, however brief that moment may be. Each moment the particle having its own wavelength, wave-phase, frequency (or energy), etc.

Because of the gravitational properties of the empty space in which these little quanta oscillated, they grew into ever more complex wave structures, eventually forming three dimensional structures, in spacetime. So, what we now have are these tiny little particles following information that exists embedded into their own particular matter waves. But these particles kept growing in complexity, eventually becoming human beings. Remember, information begets information, it is its nature. In my view, our consciousness is represented by this pilot wave, which is still present and of the utmost importance in the development of each one of us.

Sentience, in this view, is analogous to active information. Just as cellular automata can produce infinite complexity from only a few laws, active information follows Nature's four fundamental forces, which are directed at the aether level, constantly recreating itself according to information already contained within each particular system's pilot wave and the environment in which it evolves. Sentience, the way I see it, comes from an inherent self-reference mechanism in active information.

Sentience is probably a set of built-in algorithms... archetypes... memes... like beautiful passages in a musical composition integrated into biological matter's morphic waves whose functions are to make information grow, self-organize, replicate, and preserve itself, all at the same time. It is active information. Do these algorithms come from Natural selection? I see this information gathering function in self-animated matter as a representation of a process ruled by a syntropic principle. The Mandelbrot set comes to mind.

One could see Nature, in this sense, driven by a hunger, an urge to continue to exist until it satisfies a need to balance and harmonize. An urge to order reality; to gather and preserve the information that guarantees its long term survival. It is what makes us continue, it is what makes all matter continue. Could this be the reason why biological matter seems to be following a preferred time line?

As long as the basic laws remain the same, the universe will always anthropomorphize itself in the sense that, whatever comes out as a top product will likely have many of the qualities characteristic to human beings. We were not created in God's image, the universe created itself in Man's image. After dinosaurs disappeared, humanoids were the most likely outcome on this planet. Make any changes to any of the fundamental forces, and

90

we may not even get a universe. But as long as the laws remain the same, we will probably end up with something similar to what we now have. It could happen anywhere in the universe, but sooner or later, as long as there is matter, it will happen. It is the way of Nature.

We are just biological units which active information employs to see the world, to exist. Look at birds, all they want in life is to get somewhere where they can mate, have offspring, provide for them, and then die. Same with salmon; in their world, getting to the top of that mountain is all that counts. To them, that is what life is all about, that is the information world they live in. A pigeon's life story may be a beautiful thing, but that is not enough, not enough information compared to what we humans can gather. As far as we know, we get God (the universe) the best possible input. Because of the way we see the world and our ability to process the information we gather through our senses, we are God's most efficient source of usable, valuable, and especially meaningful information in this neighborhood. In this sense, we truly are God's servants.

Spirit & Soul

In his book, *Answer to Job*, Carl Jung wrote:

> *"The importance of consciousness is so great that one cannot help suspecting the element of meaning to be concealed somewhere within all the monstrous, apparently senseless biological turmoil, and that the*

91

*road to its manifestation was ultimately
found on the level of warm-blooded
vertebrates possessed of a differentiated
brain - found as if by chance, unintended and
unforeseen, and yet somehow sensed, felt and
groped for out of some dark urge."*

Spirit is one. You may say, but how am I connected to
God? How can God be every man, or every man be God?
Well, that is why I am obsessed with the concept of an
aether, or empty space, if you prefer to call it that.
Remember, thanks to wholeness, state, not knowledge,
instantaneously spreads throughout the whole universe.
Empty space is all pervading, it is the space between the
points. And the points, the particles, are just clusters,
nodes of information floating in that empty space, as a
hologram ruled by the laws of Quantum Mechanics,
Thermodynamics, and Relativity. Because this empty
space permeates everything, it is omnipresent. So, it does
not matter where you are, you could be a man in Buenos
Aires, Montreal, or Tokyo... you would still be connected
to God through the empty space in which the universe
sits. As David Bohm said, empty space is not what
separates us, it is what unites us!

We are basically a collection of cells, an organism,
moving around in 3D space. That is why we need eyes.
When you need something, you locate it using your eyes,
then you lift an arm and reach for it. When you walk, you
first look for a safe path, then start walking through it.
But, who's the one looking? What is the Self? When you
introspect, who's the one doing the introspection? When

we say I, who do we really mean? We know we are the ones thinking, Descartes got that right. And we also know that we are a bunch of cells floating in an empty space that permeates everything.

There is a universal being that we are connected to, then there is our soul to which we are also connected, and which serves as our own particular filter. The soul is a compilation of all of the experiences we have had as we interacted with the world. It is the lens through which we as individuals see the world. In my view, it is a field, or what some call, a unified consciousness field. In this view, fields are material, but the universal being we are connected to, is not. The collective unconscious, as Jung used to call it, is also a material field, a compilation of all the experiences our species has had, each species having its own field. Teilhard de Chardin called it the Noosphere, and it is supposed to surround planet Earth.

Purpose

We cannot use subjectivity to understand or explain objective reality. Love, purpose... these things have meaning to us, but neither Science nor the universe cares about them. To understand and explain reality, we must look at it in an objective way, and the facts say that the aether... God... Mind... Consciousness... the Self... or whatever, is just a thing with no purpose. It can neither see, nor think... until process turns it into brains with eyes. We see and think for it, and we like it and want to continue, that is where purpose comes from. We have purpose, the aether does not. In other words, the quantum state of the universe is barely affected by the quantum state of Humanity, or planet Earth. Its state depends on many other factors. Also, as far as we know, many other extraterrestrial civilizations may be also contributing to the universe's overall state.

Material systems will use only that information which contains meaning to them, or meaningful information. Whatever happens here on Earth has meaning only to us humans, and we, as Nature's top product, are accountable for it. Scientific research is our obligation. By the same token, whatever advances in science extraterrestrial beings may be responsible for, would never be transferred to us... unless we went to one of their schools.

In this part of the universe, humans, after 13.7 billion years of evolution, are state of the art. Our brain, Nature's jewel in the crown, will be doing what it created itself to do: which is to gather the best available information in order to better understand and enjoy existence. Always thriving to maintain a state of well being, stability, and tranquility.

Will

Why does matter try to better itself, where does this syntropy come from? Is the best possible choice always selected? I tend to think it is a phenomenon ruled by the laws of Thermodynamics, the path of less resistance, or the most energy efficient, meaningful, and useful process. But at the same time, I see the aether as Spirit, and the place where energy and Will come from. Will that manifests itself as an urge to fight entropy, a need to order, balance, stabilize, and preserve information, e.g., atoms and DNA. Maybe, this is where Humanity's hunger for knowledge comes from.

When your thoughts wander away, what brings them back, what makes you pay attention? Will does, that is where intention comes from, but this thing we call aether can neither think nor see by itself, it needs our brains. There is Will, energy, and matter, then come human consciousness and purpose.

As I said before, all you need to be physical is to be able to act. *To will is to act.* There is matter because there is Will. This is where active information, or quanta come from. Quanta is defined as a quantum of action, in other words, an amount of energy. Or, a quantized amount of fluctuating spacetime.

Will is intrinsic to matter because matter sits on the aether. Energy, or Will without purpose, is but does not exist... until it turns into spacetime; purpose comes afterward.

Aether/God/Spirit -> Energy/Will -> Matter/Information/Purpose

Being

If we are to answer -- *What is that which is?* -- we need the notion of an aether. After all, what is real, this ever changing material reality, or the eternal? What was five seconds ago, is no longer.

The aether is that which *is*. It is immutable, it is now what it always was, and simply because matter is in constant change, there is nothing in this universe you can say that about.

That which is needs to be something eternal, immutable, and absolute, with no beginning and no ending. For that to be possible there needs to be no motion, therefore no time. Matter is then ruled out. Matter is what it is only at present time, neither the past, nor the future exist as matter. Whether you call it God or not, would depend on what you think God is. To me, God is a thing, an entity incapable of thinking until matter and brains come to existence. Many call it Cosmic Consciousness, others call it Mind, or the Self, but they are all referring to the same thing: a universal being. The aether, like God, is omnipresent and eternal, with no beginning and no ending. The aether is the seat to all fields, and without fields there can be no universe, therefore it is the source to everything there is.

God (aether) is Spirit, Will without purpose... pure energy, which is neither hot, nor bright. *Aithor* means *I burn*, maybe that is why the meaning of the word *aether* is *maker*, or *burner*, as in *the fire that builds*. [Thousands of years ago, Aether and Thor were also known, respectively, as the gods of light and thunder.]

Thanks to the aether, state, not information, instantaneously spreads throughout the universe. God feels what biological matter feels, and the day it feels at peace and fulfilled, or the day there is no more uncertainty, will be the day the universe freezes and goes back to being just flat empty space, but in a different state of being.

Aether is the physicalists' god.

Quotes and Excerpts

"That gravity should be innate, inherent, and essential to matter, so that one body may act upon another at a distance through a vacuum, without the mediation of anything else, and by and through which their action and force may be conveyed from one to another, is to me so great an absurdity that I believe no man who has in philosophical matters a competent faculty of thinking can ever fall into it. Gravity must be caused by an agent acting constantly according to certain laws, but whether this agent be material or immaterial I have left to the consideration of my readers."

"I here use the word attraction for any endeavor whatever made by bodies to approach each other; whether that endeavor arise from the action of the bodies themselves as tending mutually to, or agitating each other by spirits emitted; or whether it arises from the action of the ether or of the air or of any medium whatever [...] upon as real, to enable acceleration or rotation to be looked upon as something real." --- Isaac Newton (1693)

"Absolute Space in its own nature, without relation to anything external remains always similar and immovable." --- Isaac Newton (1687)

"I agree with you that the general relativity theory admits of an ether hypothesis as does the special relativity theory. But this new ether theory would not violate the principle of relativity. The reason is that the state [...metric tensor] = Aether is not that of a rigid body in an independent state of motion, but a state of motion which is a function of position determined through the metrical phenomena." --- Albert Einstein (1916)

"But on the other hand there is a weighty argument to be adduced in favour of the ether hypothesis. To deny the ether is ultimately to assume that empty space has no physical qualities whatever.

[...]

Recapitulating, we may say that according to the general theory of relativity space is endowed with physical qualities; in this sense, therefore, there exists an ether. According to the general theory of relativity space without ether is unthinkable; for in such space there not only would be no propagation of light, but also no possibility of existence for standards of space and time (measuring-rods and clocks), nor therefore any spacetime intervals in the physical sense. But this ether may not be thought of as endowed with the quality characteristic of ponderable media, as consisting of parts which may be tracked through time. The idea of motion may not be applied to it." --- Albert Einstein (*Ether and the Theory of Relativity*, 1920)

"The inseparability of time and space emerged in connection with electrodynamics, or the law of propagation of light. With the discovery of the relativity of simultaneity, space and time were merged in a single continuum in a way similar to that in which the three dimensions of space had previously merged into a single continuum. Physical space was thus extended to a four dimensional space which also included the dimension of time. The four dimensional space of the special theory of relativity is just as rigid and absolute as Newton's space." --- Albert Einstein (1954)

"When forced to summarize the general theory of relativity in one sentence: Time and space and gravitation have no separate existence from matter. ... Physical objects are not in space, but these objects are spatially extended. In this way the concept 'empty space' loses its meaning. ... The particle can only appear as a limited region in space in which the field strength or the energy density are particularly high..." --- Albert Einstein

---

"Nobody would believe that the chance disturbance--say by an impact--of one body in a system of uninfluenced bodies which are left to themselves and move uniformly in a straight line, where all the bodies combine to fix the system of coordinates, will immediately cause a disturbance of the others as a consequence.

We should [...] have to picture to ourselves some other medium, filling, say, all space, with respect to the constitution of which and its kinetic relations to the bodies placed in it we have at present no adequate knowledge. In itself such a state of things would not belong to the impossibilities. [...] we might still hope to learn more in the future concerning this hypothetical medium; and from the point of view of science it would be in every respect a more valuable acquisition than the forlorn idea of absolute space." --- Ernst Mach (1893)

"Having recognized that the individual points in Newton's absolute space have no physical reality, we must now inquire what remains of this concept at all." --- Max Born

----

"I cannot but regard the ether, which can be the seat of an electromagnetic field with its energy and its vibrations, as endowed with a certain degree of substantiality, however different it may be from all ordinary matter." --- Hendrik Lorentz (1906)

"Natural science (physics) contains in itself synthetical judgments a priori, as principles. ... Space then is a necessary representation a priori, which serves for the foundation of all external intuitions." --- Immanuel Kant (1781)

"I cannot conceive curved lines of force without the conditions of a physical existence in that intermediate space." --- Faraday (1830)

"...quantum field theory says that associated with any mass m there is a length called its Compton wavelength, lc, such that determining the position of a particle of mass m to within one Compton wavelength requires enough energy to create another particle of that mass. Particle creation is a quintessentially quantum-field-theoretic phenomenon. Thus, we may say that the Compton wavelength sets the distance scale at which quantum field theory becomes crucial for understanding the behaviour of a particle of a given mass. On the other hand, general relativity says that associated to any mass m there is a length called the Schwarzschild radius, ls, such that compressing an object of mass m to a size smaller than this results in the formation of a black hole. The Schwarzschild radius is roughly the distance scale at which general relativity becomes crucial for understanding the behaviour of a given mass. Now, ignoring some numerical factors, we have:

$lc = hbar/mc$

and

$ls = Gm/c^2$

These two lengths become equal when m is the Planck

mass. And when this happens, they both equal the Planck length!"

[...]

"...in topological quantum field theory we cannot measure time in seconds, because there is no background metric available to let us count the passage of time! We can only keep track of topological change."

"The topology of spacetime is arbitrary and there is no background metric."

"Quantum topology is very technical, as anything involving mathematical physicists inevitably becomes. But if we stand back a moment, it should be perfectly obvious that differential topology and quantum theory must merge if we are to understand background-free quantum field theories. In physics that ignores general relativity, we treat space as a background on which the process of change occurs. But these are idealizations which we must overcome in a background-free theory. In fact, the concepts of 'space' and 'state' are two aspects of a unified whole, and likewise for the concepts of 'spacetime' and 'process'. It is a challenge, not just for mathematical physicists, but also for philosophers, to understand this more deeply." --- John C. Baez (from *Higher-dimensional algebra and Planck scale physics*, as it appeared on the book *Physics Meets Philosophy at the Planck Scale* by Craig Callender and Nick Hugget)

"Quantum phenomena are caused by fractal topological

defects embedded in and forming a growing three-dimensional fractal process-space, which is essentially a quantum foam." --- Reginald T. Cahill

About Process Physics:

...The guiding idea of its approach is that natural existence consists in and is best understood in terms of processes rather than things--of modes of change rather than fixed stabilities. For processists, change of every sort--physical, organic, psychological--is the pervasive and predominant feature of the real.

Process philosophy diametrically opposes the view--as old as Parmenides and Zeno and the Atomists of Pre-Socratic Greece--that denies processes or downgrades them in the order of being or of understanding by subordinating them to substantial things. By contrast, process philosophy pivots on the thesis that the processual nature of existence is a fundamental fact with which any adequate metaphysic must come to terms.

Process philosophy puts processes at the forefront of philosophical and specifically of ontological concern. Process should here be construed in pretty much the usual way--as a sequentially structured sequence of successive stages or phases...

http://plato.stanford.edu/entries/process-philosophy/

[...]

Process philosophy is a longstanding philosophical

tradition that emphasizes becoming and changing over static being...

...the world, at its most fundamental level, is made up of momentary events of experience rather than enduring material substances. Process philosophy speculates that these momentary events, called "actual occasions" or "actual entities," are essentially self-determining, experiential, and internally related to each other.

Actual occasions correspond to electrons and sub-atomic particles, but also to human persons. The human person is a society of billions of these occasions (that is, the body), which is organized and coordinated by a single dominant occasion (that is, the mind). Thus, process philosophy avoids a strict mind-body dualism....

http://en.wikipedia.org/wiki/Process_philosophy

http://plato.stanford.edu/entries/spacetime-bebecome/

"When theorizing about an all-inclusive reality, the first and most important principle is containment, which simply tells us what we should and should not be considering. Containment principles, already well known in cosmology, generally take the form of tautologies; e.g., "The physical universe contains all and only that which is physical." The predicate "physical", like all predicates, here corresponds to a structured set, "the physical universe" (because the universe has structure and contains objects, it is a structured set). But this usage of tautology is somewhat loose, for it technically amounts to a predicate-logical equivalent of propositional tautology called autology, meaning self-description. Specifically, the predicate physical is being defined on topological containment in the physical universe, which is tacitly defined on and descriptively contained in the predicate physical, so that the self-definition of "physical" is a two-step operation involving both topological and descriptive containment. While this principle, which we might regard as a statement of "physicalism", is often confused with materialism on the grounds that "physical" equals "material", the material may in fact be only a part of what makes up the physical. Similarly, the physical may only be a part of what makes up the real. Because the content of reality is a matter of science as opposed to mere semantics, this issue can be resolved only by rational or empirical evidence, not by assumption alone." --- Christopher Michael Langan

http://www.ctmu.org/Articles/IntroCTMU.htm

"As everyone knows, the aether played a great part in the physics of the nineteenth century; but in the first decade of the twentieth, chiefly as result of the failure of attempts to observe the earth's motion relative to the aether, and the acceptance of the principle that such attempts must always fail, the word "aether" fell out of favour, and it became customary to refer to the interplanetary spaces as "vacuous"; the vacuum being conceived as mere emptiness, having no properties except that of propagating electromagnetic waves. But with the development of quantum electrodynamics, the vacuum has come to be regarded as the seat of the "zero-point" oscillations of the electromagnetic field, of the "zero-point" fluctuations of electric charge and current, and of a "polarization" corresponding to a dielectric constant different from unity. It seems absurd to retain the name "vacuum" for an entity so rich in physical properties, and the historical word "aether" may fitly be retained." --- Sir Edmund T. Whittaker (in the preface to his scholarly and scientific *A History of the Theories of Aether and Electricity*, 1951)

"The aetherless basis of physical theory may have reached the end of its capabilities and we see in the aether a new hope for the future." --- Paul A.M. Dirac (Physics Nobel Prize winner, 1954)

[...]

"Physical knowledge has advanced much since 1905, notably by the arrival of quantum mechanics, and the situation [about the scientific plausibility of aether] has again changed. If one examines the question in the light of present-day knowledge, one finds that the aether is no longer ruled out by relativity, and good reasons can now be advanced for postulating an aether... We can now see that we may very well have an aether, subject to quantum mechanics and conformable to relativity, provided we are willing to consider a perfect vacuum as an idealized state, not attainable in practice. From the experimental point of view there does not seem to be any objection to this. We must make some profound alterations to the theoretical idea of the vacuum... Thus, with the new theory of electrodynamics we are rather forced to have an aether." -

-- Paul A.M.Dirac (*Is There an Aether?*, Nature 168 (1951): 906-7)

In an article on *Ether* for the Encyclopedia Britannica Maxwell wrote:

"Ether or Aether ('aiqhr', probably from 'aiqw' - I burn) a material substance of a more subtle kind than visible bodies, supposed to exist in those parts of space which are apparently empty... Whatever difficulties we may have in forming a consistent idea of the constitution of the aether, there can be no doubt that the interplanetary and interstellar spaces are not empty, but are occupied by a material substance or body, which is certainly the largest, and probably the most uniform body of which we have any knowledge. Whether this vast homogeneous expanse of isotropic matter is fitted not only to be a medium of physical interaction between distant bodies, and to fulfill other physical functions of which, perhaps, we have as yet no conception, but also ... to constitute the material organism of beings exercising functions of life and mind as high or higher than ours are at present - is a question far transcending the limits of physical speculation."

[...]

Maxwell's words at the conclusion of his 1873 work on electricity and magnetism:

"All these theories lead to the conception of a medium..., and if we admit this medium as a hypothesis, I think it ought to occupy a prominent place in our investigations, and that we ought to endeavor to construct a mental representation of all the details of its actions, and this has been my constant aim in this treatise."

http://www.mathpages.com/home/kmath322/kmath322.htm

"Well, perhaps we should finish with this business about empty space.

If you follow through the mathematics of the present Quantum Theory, it treats the particle as what is called the quantized state of the field, that is, as a field spread over space but in some mysterious way with a quantum of energy. Now each wave in the field has a certain quantum of energy proportional to its frequency. And if you take the electromagnetic field, for example, in empty space, every wave has what is called a zero point energy below which it cannot go, even when there is no energy available. If you were to add up all the waves in any region of empty space you would find that they have an infinite amount of energy because an infinite number of waves are possible. Now, however, you may have reason to suppose that the energy may not be infinite, that maybe you cannot keep on adding waves that are shorter and shorter, each contributing to the energy. There may be some shortest possible wave, and then the total number of waves would be finite and the energy would also be finite. Now, you have to ask what would be the shortest length and there seems to be reason to suspect that the gravitational theory may provide us with some shortest length, for according to general relativity, the gravitational field also determines what is meant by "length" and metric. If you said the gravitational field was made up of waves which were quantized in this way, you would find that there was a certain length below which the gravitational field would become undefinable because of this zero point movement and you wouldn't be able to define length. Therefore, you could say the property of measurement, length, fades out at very short distance and you'd find the place at which it fades out would be about 10^{-33} cm. That is a very short distance because the shortest distances that physicists have ever probed so far might be 10^{-16} cm. or so, and that's a long way to go. If you then compute the amount of energy that would be in space, with that shortest possible wave length, then it turns out that the energy in one cubic centimeter would be immensely beyond the total energy of all the known matter in the universe.

Present theory says that the vacuum contains all this energy which is then ignored because it cannot be measured by an instrument. The philosophy being that only what could be measured by an instrument could be considered to be real, because the only point about the reality of physics is the result of instruments, except that it is also said that there are particles there that cannot be seen in instruments at all. What you can say is that the present state of theoretical physics implies that empty space has all this energy, and matter is a slight increase of the energy, and therefore matter is like a small ripple on this tremendous ocean of energy, having some relative stability, and being manifest. Now, therefore, my suggestion is that this implicate order implies a reality immensely beyond what we call matter. Matter itself is merely a ripple in this background.

If you take a crystal which is at absolute zero it does not scatter electrons. They go through it as if it were empty. And as soon as you raise the temperature and (produce) inhomogeneities, they scatter. Now, if you used those electrons to observe the crystal (e.g., by focusing them with an electron lens to make an image), all you would see would be these little inhomogeneities and you would say they are what exists and the crystal is what does not exist. Right? I think this is a familiar idea, namely to say that what we see immediately is really a very superficial affair. However, the positivist used to say that what we see immediately is all there is or all that counts, and that our ideas must simply correlate what we see immediately.

So now, with this vast reserve of energy and empty space, saying that matter itself is that small wave on empty space, then we could better say that the space as a whole (and we start from the general space) is the ground of existence, and we are in it. So the space doesn't separate us, it unites us. Therefore it's like saying that there are two separate points and a certain dotted line connects them, which shows how we think they are related, or to say there is a real line and that the points are abstractions from that.

The line is the reality and the points are abstractions. In that sense we say that there are no separate people, you see, but that 'that' is an abstraction which comes by taking certain features as abstracted and self-existent." --- David Bohm (*Wholeness and the Implicate Order*)

"The fundamental element of the cosmos is Space. Space is the all-embracing principle of higher unity. Nothing can exist without Space. .. According to ancient Indian tradition the Universe reveals itself in two fundamental properties: as Motion and as that in which motion takes place, namely Space. This Space is called Akasa .. derived from the root kas, 'to radiate, to shine', and has therefore the meaning of ether which is conceived as the medium of movement. The principle of movement, however, is Prana, the breath of life, the all-powerful, all-pervading rhythm of the universe. Space is the precondition of all that exists, be it material or immaterial form, because we can neither imagine an object nor a being without space. Space, therefore, is not only a conditio sine qua non of all existence, but a fundamental property of our consciousness. Our consciousness determines the kind of space in which we live. The infinity of space and the infinity of consciousness are identical. In the moment in which a being becomes conscious of his consciousness, he becomes conscious of space. In the moment in which he becomes conscious of the infinity of space, he realises the infinity of consciousness." --- Lama Anagarika Govinda (1969)

"There is a thing, formless yet complete. Before heaven and earth it existed. Without sound, without substance, it stands alone and unchanging. It is all-pervading and unfailing. One may think of it as the mother of all beneath Heaven. We do not know its name, but we call it Tao. Deep and still, it seems to have existed forever." ---

Lao Tzu

Book and Article Resources

Aharonov Y. and Suhail Zubairy M. - Time and the Quantum: Erasing the Past and Impacting the Future, Science 11 February 2005: 875-879

Aspect, A., Grangier, P., and Roger, G. (1982) Experimental realization of Einstein-Podolsky-Rosen-Bohm Gedankenexperiment: a new violation of Bell's inequalities. Phys. Rev. Lett. 48:91-94.

Beck, F. and Eccles, J.C. (1992) Quantum aspects of brain activity and the role of consciousness. Proc. Natl. Acad. Sci. USA 89(23):1135711361.

Bekenstein, J.D. Information in the Holographic Universe. Scientific American, Volume 289, Number 2, August 2003, p. 61.

Bekenstein, J.D. and Schiffer, M. (1990) "Quantum Limitations on the Storage and Transmission of Information", Int. J. of Modern Physics 1:355-422

Bekenstein, J.D. (1984) "Entropy content and information

flow in systems with limited energy", Phys. Rev. D 30:1669–1679

Bell, J.S. (1987) Speakable and unspeakable in quantum mechanics: Collected papers on quantum philosophy. Cambridge, Cambridge University Press.

BELL, J.S. (1966) On the problem of hidden variables in quantum theory, Reviews of Modern Physics, 38, p. 447.

Bell, J. (1964), 'On the Einstein Podolsky Rosen Paradox', Physics, 1 (195).

Bohm, D. and Hiley B.J. (1993) The Undivided Universe. Routledge, New York

Bohm, D. Wholeness and the Implicate Order, Routledge and Kegan Paul, London (1980).

Boyer, Timothy - The Classical Vacuum [Zero Point Energy] Scientific American, Aug. 1985, pp 70-78.

Callender C., Huggett, N. (2001) Physics Meets Philosophy at the Planck Scale: Contemporary Theories in Quantum Gravity Publisher: Cambridge University Press.

Capra, F. (2000) The Tao of Physics: An Exploration of the Parallels between Modern Physics and Eastern Mysticism Publisher: Shambhala Publications, Inc.

Chalmers, David J. in Conscious Experience - Absent Qualia, Fading Qualia, Dancing Qualia. Edited by

Thomas Metzinger. Ferdinand Schoningh, 1995.

Chalmers, D.J. (1997), 'Moving forward on the problem of consciousness', JCS, 4 (1), pp. 3–46.

Chalmers, D. (1996) Facing up to the problem of consciousness. In: Toward a Science of Consciousness - The First Tucson Discussions and Debates, S.R. Hameroff, A. Kaszniak and A.C. Scott (eds.), MIT Press, Cambridge, MA.

Churchland, P.M. 1996. The rediscovery of light. Journal of Philosophy 93:211-28.

Conrad M. 1993. "The Fluctuon Model of Force, Life and Computation: A Constructive Analysis." Appl. Math. Comput. 56:203-59.

Cramer, J.G. 1986. The transactional interpretation of quantum mechanics. Review of Modern Physics 58:647-87.

Crick, F., and Koch, C. (1990) Towards a neurobiological theory of consciousness. Seminars in the Neurosciences 2, 263-275

Crick, F. and Koch, C. 1995. Why neuroscience may be able to explain consciousness. Scientific American 273(6):84-85.

Davies, P. (1998) The Fifth Miracle: The Search for the Origin of Life, Penguin, London.

Davies, P. "Physics and life" in: The First Steps of Life in the Universe Proceedings of the Sixth Trieste Conference on Chemical Evolution. Trieste, Italy, Editors: Chela-Flores, J., Owen, Tobias and Raulin, F. Kluwer Academic Publishers: Dordrecht, The Netherlands.

Dawkins, R. 1989. The selfish gene (revised edition) Oxford Press, Oxford U.K.

Dennett D.C. 1995. Darwin's dangerous idea: Evolution and the meanings of life. Touchstone, New York.

Dennett, D.C. 1996. Facing backwards on the problem of consciousness. Journal of Consciousness Studies 3:4-6.

Dirac, P. The Principles of Quantum Mechanics, 4th ed. (Oxford University Press, Oxford, UK, 1958).

Einstein, A. The Meaning of Relativity. Princeton University Press. Princeton, N. J., 1956.

Einstein's opinions of Bohm's work reported in M. Talbot, The Holographic Universe (New York: HarperCollins, 1991), 39.

Fontana,W., [1991]. "Algorithmic Chemistry". In: Artificial Life II. Langton,C.G., Taylor,C., Farmer,J.D.,Rasmussen,S. (Eds.). Addison-Wesley, pp. 159-209.

Frohlich H (1970) Long range coherence and the actions of enzymes. Nature 228:1093.

Gabor, D. "Holography, 1948-1971." Science, vol. 177, pp. 299-313, 1972.

Hameroff SR, and Penrose R (1996b) Conscious events as orchestrated spacetime selections. Journal of Consciousness Studies 3(1):36-53 http: //www.u.arizona.edu/~hameroff/penrose2

Hameroff S.R. and Penrose R. (1996a) Orchestrated reduction of quantum coherence in brain microtubules: A model for consciousness. In: Toward a Science of Consciousness - The First Tucson Discussions and Debates, S.R. Hameroff, A. Kaszniak and A.C. Scott (eds.), MIT Press, Cambridge, MA, pp. 507-540

Hameroff S.R. (1998) Funda-mentality: is the conscious mind subtly linked to a basic level of the universe? Trends in Cognitive Science 2(4)119-127

Hameroff S.R. and Watt R.C. (1983) Do anesthetics act by altering electron mobility? Anesth. Analg. 62, 936-940.

Hameroff, S.R. 1994. Quantum coherence in microtubules: A neural basis for emergent consciousness? Journal of Consciousness Studies 1:91-118.

Hameroff, S (1998b) Did consciousness cause the Cambrian evolutionary explosion? In: Toward a Science of Consciousness II The Second Tucson Discussions and Debates. Eds S Hameroff, A Kaszniak, A Scott. MIT Press, Cambridge MA pp 421-437

Hameroff, S (1998c) "More neural than thou": Reply to Churchland's "Brainshy" in: Toward a Science of Consciousness II The Second Tucson Discussions and Debates. Eds S Hameroff, A Kaszniak, A Scott. MIT Press, Cambridge MA pp 197-213

Hawking, S. 1988. A Brief History of Time. Bantam Books.

Hiley, B. (1991), "Vacuum or Holomovement". In Philosophy of Vacuum.

James, W. (1890/1918), The Principles of Psychology, Vol. I (Henry Holt & Co.: New York).

Jibu M, Hagan S, Hameroff SR, Pribram KH & Yasue K. 1994. "Quantum Optical Coherence in Cytoskeletal Microtubules: Implications for Brain Function." Biosystems.. 32:195-209.

Jibu, M. and Yasue, K. (1997), "What is mind? Quantum field theory of evanescent photons in brain as quantum theory of consciousness", Informatica 21, pp. 471-490.

Jibu, M., Hagan, S., Hameroff, S.R., Pribram, K.H., and Yasue, K. (1994) Quantum optical coherence in cytoskeletal microtubules: implications for brain function. BioSystems 32:195209.

Kauffman Stuart (1996) At Home in the Universe: The Search for the Laws of Self-Organization and Complexity - Oxford University Press, USA

King, M. B. (1989) Tapping the Zero Point Energy
Publisher: Paraclete Pub. Co.

Kostro, Ludwik, (2000) Einstein and the Ether, Apeiron
(ISBN 0-9683689-4-8)

Lee Smolin, "Did the Universe Evolve?" Classical and
Quantum Gravity 9 (1992), pp. 173-192

Libet, B. 1996. Solutions to the hard problem of
consciousness. Journal of Consciousness Studies 3:33-35.

Lockwood, M. (1989) Mind, Brain and the Quantum
(Oxford, Basil Blackwell).

Maldacena, J., and Susskind, L. (2013), Cool horizons for
entangled black holes.
https://selectedpapers.net/arxiv/1306.0533

Maturana, H. R. & Varela, F. J. (1980), Autopoiesis and
cognition: The realization of the living (Reidel).

Maturana, H. R. & Varela, F. J. (1987), The tree of
knowledge: The biological roots of human understanding
(Shambhala Press).

McCaskill, J. S. , Tangen, U. & Ackermann, J. (1997) in
Proceedings of the 4th European Conference on Artificial
Life, eds. Husbands, P. & Harvey, I. (MIT Press/Bradford
Books, Cambridge, MA), pp. 398-406.

Penrose, R. (1989) The Emperor's New Mind: concerning
minds, computers and the laws of physics (Oxford,

Oxford University Press).

Penrose, R. (1994) Shadows of the Mind. Oxford: Oxford University Press.

Penrose R, and Hameroff SR (1995) What gaps? Reply to Grush and Churchland. Journal of Consciousness Studies 2(2):99-112.

Pietsch, P. "Shuffle Brain." In Human Connection and the New Media, edited by B. N. Schwartz. Prentice-Hall, Englewood Cliffs, N.J., 1973.

Pietsch, P., and C. W. Schneider. "Brain Transplantation in Salamanders: An Approach to Memory Transfer." Brain Research, vol. 14, pp. 707-715, 1969.

Pribram, K. H. "The Neurophysiology of Remembering." Scientific American, January 1969.

Pribram, K. H. "The Brain." Psychology Today, September 1971a.

Pribram, K.H. (1987) "The Implicate Brain". In B.J. Hiley & F.D. Peat (eds).

Prigogine, I. (1997) The End of Certainty: Time's Flow and the Laws of Nature, Publisher: Simon & Schuster.

Popper, K.R. (1982) Quantum Theory and the Schism in Physics. London, Hutchinson.

Rosenberg, G.H. 1996. Rethinking nature: A hard

problem within the hard problem. Journal of
Consciousness Studies 3:76-88.

Russell, B. The Problems of Philosophy. Oxford
University Press, London, 1959.

Searle, J.R. (1997), The Mystery of Consciousness (New
York: New York Review of Books).

Smolin, L. (1997) Life of the Cosmos, Oxford Press,
N.Y.

Stapp, H.P. (1990) "A Quantum Theory of The Mind-
Brain Interface".

Stapp, H.P. (1991) "Quantum Propensities and the Brain-
Mind Connections".

Stapp, H.P. - Attention, Intention, and Will in Quantum
Physics - Lawrence Berkeley National Laboratory,
University of California, Berkeley, California

Stapp, H. P. (1993), Mind, Matter, and Quantum
Mechanics, Springer-Verlag, New York, Berlin,
Heidelberg.

Stapp, H. P. (1972) The Copenhagen Interpretation.
American Journal of Physics, 40, 1098-1116. [Reprinted
in Stapp (1993).]

Stapp, H. P. (1995b) Quantum Mechanical Coherence,
Resonance, and Mind. In P. R. Masini and A. Mandrekar
(Eds.), Norbert Wiener Centenary Congress, to be

published by the American Mathematical Society, New York. (Lawrence Berkeley Laboratory Report LBL-36915). See also PSYCHE (in press).

Strogatz Steven H. (2004) Sync: How Order Emerges from Chaos in the Universe, Nature, and Daily Life - Hyperion

Toffoli, T. & Margolus, N. (1987) Cellular Automata Machines (MIT Press, Cambridge, MA), pp. 119-139.

Varela, F.J. (1996), Neurophenomenology, Journal of Consciousness Studies, 3 (4), pp. 330–49.

Varela, F. 1995. Neurophenomenology: A methodological remedy for the hard problem. Journal of Consciousness Studies 3.

Weiner, N. (1936) The Role of the Observer. Philosophical Sciences 3, 307-319.

Wilber, K. (ed.) 1982. The holographic paradigm and other paradoxes. Boston: Shambhala.

Whitehead, A.N., (1929) Process and Reality. Macmillan, N.Y.

Zukav, G. (1979) The Dancing Wu Li Master: an overview of the new physics (New York, Morrow).

On-line Resources

1. Ether and the Theory of Relativity - Albert Einstein
An Address delivered on May 5th, 1920, in the
University of Leyden
http://en.wikisource.org/wiki/Ether_and_the_Theory_of_
Relativity

2. Einstein's Ether:
Why did Einstein Come Back to the Ether? by Galina
Granek
Department of Philosophy, Haifa University Mount
Carmel, Haifa 31905, Israel
http://redshift.vif.com/JournalFiles/V08NO3PDF/V08N3
GRF.PDF

3. AETHER, or Ether (Gr. *aither*, probably from aitho, I
burn, though Plato in his *Cratylus* (410 B) derives the
name from its perpetual motion-6n del *aei* thei *peri ton
aera reon, aeitheer* dikaios an kaloito), a material
substance of a more subtle kind than visible bodies,
supposed to exist in those parts of space which are
apparently empty.
http://www.1902encyclopedia.com/E/ETH/ether.html

4. Aether, Ether (Greek) (from aitho shining, fire)
The upper or purer air as opposed to aer, the lower air;
the clear sky; the abode of the gods. In Classical antiquity
it denoted primordial substance, Proteus or protyle, the

unitary source both of all substances and energies, the mask of all kosmic phenomena.
http://nicedefinition.com/Definition/Word/Aether/Aether.aspx

5. Theosophy, Vol. 50, No. 3, January, 1962 (Pages 129-134; Size: 17K)
PROTEUS(1) THE ancients called primordial Substance "Chaos." Plato and Pythagoras named it the Soul of the World. "The Mundane God, eternal, boundless, young and old, of winding form," says the Chaldean oracles. All the ancient nations deified Æther in its imponderable aspect and potency.
http://www.wisdomworld.org/additional/ListOfCollatedArticles/Proteus.html

6. The California Institute for Physics and Astrophysics (CIPA) is dedicated to exploring fundamental problems in physics (e.g. gravitation, inertia, the nature of mass) as well as very-long range technological possibilities that may emerge from the properties of the quantum vacuum.
http://www.calphysics.org/index.html

7. The Center for Consciousness Studies at the University of Arizona promotes open, rigorous discussion of all phenomena related to conscious experience. The Center organizes conferences, on-campus lectures, web courses, and hosts visiting researchers.
http://www.consciousness.arizona.edu/

8. Crisis in Life Sciences. The Wave Genetics Response. Traditionally, genetics talks about DNA, RNA and proteins' speech and texts only. The standard linguistic

structures of genome are realized at the material level in the form of sequences of "chemical letters" in a DNA chain consisting of the 2% coding DNA. In Wave Genetics the texts are realized at the material level in the form of sophisticated dynamic holograms (gene-holograms) in liquid crystals of the chromosome continuum.
http://www.emergentmind.org/gariaev06.htm

9. SHUFFLEBRAIN - The Quest of Hologramic Mind
http://www.instinct.org/texts/shufflebrain/shufflebrain-book00.html

10. PSYCHE (ISSN: 1039-723X) is a refereed electronic journal dedicated to supporting the interdisciplinary exploration of the nature of consciousness and its relation to the brain.
http://www.theassc.org/

11. If people frequently asked any questions about the CMB, then these might be among them!
http://www.astro.ubc.ca/people/scott/faq_basic.html

12. What Is the Cosmic Microwave Background Radiation?

13. Papers on Consciousness (David Chalmers).
http://consc.net/consc-papers.html

14. This site has been set up as an open venture between scientists, scholars, meditators and all those who believe that we are approaching a conceptual threshold in our

understanding of how physics, physiology and consciousness interact.
http://www.emergentmind.org/

15. Sorry Einstein, the universe needs quantum uncertainty
http://www.newscientist.com/article/mg21428702.100-sorry-einstein-the-universe-needs-quantum-uncertainty.html

16. Henry Stapp papers on consciousness
http://www-physics.lbl.gov/~stapp/stappfiles.html

17. Creation Ex Nihilo... or from Wheeler's "Pregeometry"?
http://www.quanta-gaia.org/dobson/wheelerPregeometry.html

18. Modern Scientific Theories of the ancient Aether
http://www.mountainman.com.au/aetherqr.htm

19. Studies on Consciousness, Mind and Life
http://www.thymos.com/

20. Paul J. Steinhardt - Department of Physics, Princeton University, Princeton, NJ
http://feynman.princeton.edu/~steinh/

21. Quintessence - Cosmologists have proposed that a mysterious substance called quintessence can explain why our universe is accelerating. But what is it made of?
http://physicsworld.com/cws/article/print/2000/nov/01/quintessence

22. Most approaches to the problem of consciousness see the brain as a computer, with neurons and synapses acting as switches or "bits". In this view consciousness is thought to "emerge" as a novel property of complex computation.
http://www.consciousness.arizona.edu/hameroff/

23. This is an online introduction to superstring theory, which is the leading candidate for the theory of all fundamental interactions in the universe.
http://www.sukidog.com/jpierre/strings/

24. David Bohm and Birkbeck College
http://www.bbk.ac.uk/lib/about/bohm

25. The Transactional Interpretation of Quantum Mechanics - by John G. Cramer
http://mist.npl.washington.edu/npl/int_rep/tiqm/TI_toc.html

26. Relativity, Quantum Gravity and Space-time Structures
Basil Hiley's Recent Publications
http://www.bbk.ac.uk/tpru/RecentPublications.html

27. Toward a Science of Consciousness
http://www.imprint.co.uk/Tucson2000/jcsmainframe.html

28. This introductory orientation is designed to provide you some general context regarding how autopoietic theory originated, how it developed, and where it stands now. Autopoietic theory is the term I use to denote the

work of Chilean biologists Humberto R. Maturana and Francisco J. Varela (originally labeled the biology of cognition).
http://www.enolagaia.com/Tutorial1.html

29. Consciousness and Neuroscience - Francis Crick & Christof Koch
http://www.slideshare.net/njqtpie86/consciousness-neuroscience-francis-crick-christof-koch

30. Why Classical Mechanics Cannot Naturally Accommodate Consciousness but Quantum Mechanics Can
http://arxiv.org/abs/quant-ph/9502012

31. A Conversation with Physicist Brian Greene - John Fudjack
http://www.tap3x.net/EMBTI/j6greene.html

32. Quantum Teleportation
http://www.almaden.ibm.com/st/past_projects/quantum_information/

33. Holism and Nonseparability in Physics
http://plato.stanford.edu/entries/physics-holism/

34. Quantum Mechanics: 1-Dimensional Particle States Applet
This java applet is a quantum mechanics simulation that shows the behavior of a single particle in bound states in one dimension. It solves the Schrödinger equation and allows you to visualize the solutions.
http://www.falstad.com/qm1d/

35. Experiencing Soun-gui - Jean-Luc Nancy
http://www.usc.edu/dept/comp-lit/tympanum/3/nancy.html

36. On Exploring New Approaches Within Physics - John K. N. Murphy
http://www.hotquanta.com/

37. Quantum Nonlocality and the Possibility of Superluminal Effects - John G. Cramer
http://www.npl.washington.edu/npl/int_rep/qm_nl.html

38. Self-Organizing Systems (SOS) FAQ
http://www.calresco.org/sos/sosfaq.htm

39. Autopoietic Theory: Deeper Discussion
http://www.enolagaia.com/ATDefs.html

40. David Bohm and the Implicate Order - by David Pratt
http://www.theosophy-nw.org/theosnw/science/prat-boh.htm

41. It's confirmed: Matter is merely vacuum fluctuations
Matter is built on flaky foundations. Physicists have now confirmed that the apparently substantial stuff is actually no more than fluctuations in the quantum vacuum.
http://www.newscientist.com/article/dn16095-its-confirmed-matter-is-merely-vacuum-fluctuations.html

42. Comparison between Karl Pribram's "Holographic Brain Theory" and more conventional models of neuronal computation

http://www.acsa2000.net/bcngroup/jponkp/

43. Alan Guth: A Golden Age of Cosmology
Even though cosmology does not have that much to do
with information It certainly does have to do with
revolution and phase transitions, in fact phase transitions
in both the literal and the figurative sense of the word.
http://edge.org/video/a-golden-age-of-cosmology

44. Our world may be a giant hologram
"If the GEO600 result is what I suspect it is, then we are
all living in a giant cosmic hologram." --- Craig Hogan,
director of Fermilab's Center for Particle Astrophysics.
http://www.newscientist.com/article/mg20126911.300-
our-world-may-be-a-giant-hologram.html

45. Einstein and the Ether - by Ludwik Kostro (Apeiron,
Montreal, 2000)
Whether gravitational, electrical, and nuclear interactions
can be encompassed within a unified theoretical structure,
and whether such a structure will be conceived as a
plenary space with physical properties, remains to be
seen. But if the history of the successive dynasties of
aether is any guide, we can eventually proclaim:
The luminiferous aether is dead!
Long live the aether!" (Owen Gingerich)
http://itis.volta.alessandria.it/episteme/ep3-24.htm

46. Eye and Thou (Dissolving Descartes)
Capstone Address to the IEEE Visualization '97
Conference 24 October 1997
http://www.hyperreal.org/~mpesce/viz97.html

47. Gravitation as a pressure force: a scalar ether theory
http://geo.hmg.inpg.fr/arminjon/PIR96_1B.pdf

48. Problems of the Inhomogeneous Physical Vacuum
http://www.sinor.ru/~che/Vdyatlov1.htm

49. General Relativity and Spatial Flows: I. Absolute
Relativistic Dynamics
http://xxx.lanl.gov/abs/gr-qc/0006029

50. Simple, common-sense physics The mechanical
comprehension of the nature of the cosmos of the all-
pervading aether
http://www.aethro-kinematics.com/

51. Astronomy 123: Galaxies and the Expanding
Universe
http://abyss.uoregon.edu/~js/ast123/

52. Quantum Consciousness - by Piero Scaruffi

53. Flowing Space - by Henry H. Lindner
http://home.epix.net/~hhlindner/Writings/Implicate/Impli
cations.html

54. From the Heisenberg Picture to Bohm: a New
Perspective on Active Information and its relation to
Shannon Information.
http://www.bbk.ac.uk/tpru/BasilHiley/Vexjo2001W.pdf

55. Non-commutative Geometry, the Bohm Interpretation
and the

Mind-Matter Relationship - B. J. Hiley
http://www.bbk.ac.uk/tpru/BasilHiley/noncommgeobohm
.pdf

56. Foundation Reasoning of Electrogravitational Theory
and Tests
http://www.electrogravity.com/index7.html

57. Models of Self-Organization Using Genetic Cell
Automata
https://elibrary.asabe.org/abstract.asp?aid=13738&t=2&r
edir=&redirType=

58. Despite several thousand years of failure to correctly
understand physical reality (hence the current postmodern
view that this is impossible) it is actually very simple to
work out how matter exists and moves about in Space.
http://www.spaceandmotion.com/Physics-Space-Aether-
Ether.htm

59. I said, at the Aether scale there are no distances to
cover, it is everywhere. And, as Eugene V. Stefanovich
contends: interactions, not forces, are instantaneously
registered throughout space. Finally, someone offering a
good explanation for instantaneous state transfers.
http://arxiv.org/abs/physics/0612019 /
http://www.physicsforums.com/showthread.php?t=17596
5

60. A Universe From Nothing' by Lawrence Krauss,
Richard Dawkins, AAI 2009
http://www.youtube.com/watch?v=EjaGktVQdNg

61. The Mystery of Empty Space
University of California Television (UCTV)
http://www.youtube.com/watch?v=Y-vKh_jKX7Q

62. Why we don't 'need' an aether to explain Relativity
"When dimensions are understood as mere components of the grid system, rather than physical attributes of space, it is easier to understand the alternate dimensional views as being simply the result of coordinate transformations."
http://en.wikipedia.org/wiki/Spacetime

63. Gravity Being Described as Space Flow
Excerpt from Brian Cox's "Wonders of the Universe"
https://drive.google.com/file/d/0ByfoRemtwEkpQUE5M
F9xWkR6UWc/edit?usp=sharing

64. Holographic principle
http://en.wikipedia.org/wiki/Holographic_principle

65. The Quantum Fabric of Space-Time
https://www.quantamagazine.org/20150424-wormholes-entanglement-firewalls-er-epr/

66. The Good Vibrations of Quantum Field Theories
By Don Lincoln
http://www.pbs.org/wgbh/nova/blogs/physics/2013/08/the-good-vibrations-of-quantum-field-theories/

67. Relativity Experimentally Confirmed
https://drive.google.com/file/d/0ByfoRemtwEkpRUV6Q
m44d3gzV1k/view?usp=sharing

68. How Quantum Pairs Stitch Space-Time
New tools may reveal how quantum information builds the structure of space.
https://www.quantamagazine.org/20150428-how-quantum-pairs-stitch-space-time/